Lean RFS
(Repetitive Flexible Supply)

PUTTING THE PIECES TOGETHER

Ian Fraser Glenday • Rick Sather

With a Foreword by Daniel T. Jones

Lean RFS
(Repetitive Flexible Supply)

PUTTING THE PIECES TOGETHER

CRC Press
Taylor & Francis Group
Boca Raton   London   New York

CRC Press is an imprint of the
Taylor & Francis Group, an **informa** business

A PRODUCTIVITY PRESS BOOK

CRC Press
Taylor & Francis Group
6000 Broken Sound Parkway NW, Suite 300
Boca Raton, FL 33487-2742

© 2013 by Taylor & Francis Group, LLC
CRC Press is an imprint of Taylor & Francis Group, an Informa business

No claim to original U.S. Government works

Printed on acid-free paper
Version Date: 20130422

International Standard Book Number-13: 978-1-4665-7819-7 (Paperback)

Library of Congress Cataloging-in-Publication Data

Glenday, Ian.
  Lean RFS (repetitive flexible supply) : putting the pieces together / Ian Fraser Glenday, Ricky L. Sather.
    pages cm
  Includes bibliographical references and index.
  ISBN 978-1-4665-7819-7 (pbk. : alk. paper)
  1. Business logistics. 2. Lean manufacturing. 3. Organizational effectiveness. I. Sather, Ricky L. II. Title.

HD38.5.G56 2013
658.7--dc23   2013007326

Visit the Taylor & Francis Web site at
http://www.taylorandfrancis.com

and the CRC Press Web site at
http://www.crcpress.com

Contents

Foreword

Daniel T. Jones

This book is the missing link in many Lean journeys. So often, people conclude that Lean, single-piece flow will not work in their situation—whether it is making big batches of many different products through a common process, or dealing with patients with many different conditions in a hospital, or processing orders or claims requiring very different amounts of time in an office. This book shows how to begin to create flow even in these circumstances. The key insight is that a small fraction of products, patient conditions, or types of claims account for the majority of the work. Separating these from the rest, at least initially, makes it possible to create flow, to standardize and improve the way this work is done and managed. Not only will this create noticeable results quickly, but it also begins to free up time to go on to tackle the rest.

This book is also a great introduction to the power of learning by doing, which lies at the heart of Lean. Acting our way to a new way of thinking and working together is the way to challenge old mind-sets and habits. Do the *Glenday Sieve* analysis on your own work and you will be surprised how this changes the way you think. It provokes new questions that lead in quite counter-intuitive directions and takes us on a very different path down our own Lean journey. Seeing a way through the chaos and complexity of most work situations with our colleagues is the essential first step on this journey. This cannot be done for you by experts or consultants—it is a path everyone must take for himself or herself. This book will show you how.

What is also unique about this book is that it clearly spells out the theory and practice originally published in *Breaking Through to Flow*,* together with stories from Kimberly-Clark's experience in using them over many years with great success. It weaves together the experiences with the method and shows step-by-step how it builds from astonishing transformations on the shop floor to a management system higher up the organization. Kimberly-Clark and others have discovered how this thinking helps at every level, for instance in focusing actions on the vital few objectives or performance gaps in strategy formulation.

---

* *Breaking Through to Flow* by Ian Glenday, Lean Enterprise Academy, UK, 2005.

These stories give the reader a real feel of how this learning-by-doing journey led to "aha!" moments for those involved. Experienced Lean practitioners will recognize that, step-by-step, this method establishes the stability necessary to reap the benefits of the Toyota Production System in very different environments than automobile production. Indeed, it mirrors Toyota's own experiences in laying the groundwork for using TPS in batch-production environments. The steps begin by separating high- from low-volume work and establishing a pattern or rhythm to the work to expose deviations and establish stability, and they end up with the ability to align the work with both the volume and mix of demand.

I first met the coauthor of this book, Ian Glenday, while writing *Lean Thinking*.* He was working in a company making various food products in the UK. I was just getting involved in pioneering Lean supply chains with the UK supermarket chain Tesco. It puzzled me why food manufacturers glazed over when you talked about one-piece flow and synchronizing production with demand. Here was the answer.

It took several years to find willing guinea pigs, like Rick Sather (coauthor of this book) at Kimberly-Clark, to give it a go. In almost every case, Ian opened closed minds to see how this could be done, while I injected these ideas into the Efficient Consumer Response (ECR) movement in Europe. This was the missing piece of the puzzle in Leaning the fast-moving consumer goods (FMCG) industry, which is now well down the path to demand-driven, rapid-replenishment production and supply chains. Ian and I then found places to experiment with this method in all kinds of organizations, from law courts to hospitals. In each case, it produced the same initial amazement from those involved, which in turn unleashed the enthusiasm to embark on this journey. We planted many seeds that are now growing on their own accord across the world.

Ian Glenday started developing this approach in pharmaceutical and food production. He later met Yoshiki Iwata, one of the founders of the Shingijutsu Consulting Group, who at Toyota Gosei was one of the first Toyota Group suppliers to be taught the Toyota Production System by Taiichi Ohno and his team. He showed Ian the steps of leveling that Toyota used to get to one-piece-flow and encouraged Ian to develop his ideas further. This led in turn to the methodology and results described in this book.

---

* James P. Womack and Daniel T. Jones, *Lean Thinking*, Simon & Schuster, 1996.

Introduction: Lean Repetitive *flexible* Supply—Putting the Pieces Together

Repetitive *and* flexible—at the same time? Surely, that's impossible.

Using proven examples and quantifiable evidence, this book sets out to illustrate that Repetitive *flexible* Supply (R/S) can be achieved; moreover, its implementation within your company will help you reach a new level of improved performance in manufacturing and across the whole of the supply chain.

Throughout this book, Repetitive *flexible* Supply (R/S) is written with the *flexible* in italics to create an image of the principle that one can be both repetitive and *flexible* at the same time.

What Is Lean Repetitive *flexible* Supply?

The analogy we are using throughout this book for Lean/R/S is that of a jigsaw puzzle—hence our subtitle, "Putting the Pieces Together."

All of us have worked on a jigsaw puzzle at some time in our lives. Our jigsaw analogy works for us on two levels. It serves as an image of the planning process currently used by most companies; that is, of batch logic based on economically feasible order quantities. It's like every time the plan is calculated, all the pieces of the jigsaw—i.e., all the different products the company makes—are put back in the box, shaken up, and then tipped out. The planners try to put all the pieces together as best they can. The result is that all the pieces are planned every time, and—as the pieces can be assembled in many ways—one gets a different plan every time. This plan is often then changed, which can result in corporate firefighting. We find this approach endemic in most of the companies we have worked with.

When someone sets out to solve a real jigsaw puzzle, the first thing they do is find the corner pieces. Then they put in the straight edges to give shape and structure to the puzzle, thus making it easier to position the center pieces.

What would it be like if, in our planning processes, we were able to leave the corners and straight pieces in place each time we made the plan? It would create a stable, repeating structure (the repetitive products) that would make planning the center pieces (the flexible products) much easier. The manufacturer would make the key products in a repetitive pattern of the same quantities on the same day every week. This would stop a lot of unnecessary, time consuming, and costly firefighting while providing a better foundation for sustainable continuous improvement.

This is one of the key objectives of Lean/RfS—*to create a repetitive, fixed, stable plan*. Yet to many people, it seems ridiculous, impossible, and sometimes counterintuitive.

The second reason for using a jigsaw puzzle as an analogy is concerned with what we observe in most companies when they say they are applying "Lean." This usually consists of Lean tools and techniques, for example SMED, 5S, TPM, and kanbans, to name a few. When the question, "What is the main objective of Lean?" is put to people in these companies, invariably the response is, "To eliminate waste." To us, these tools and techniques are equivalent to the center pieces of a jigsaw puzzle. More importantly, however, they do not provide the outline and structure—the equivalents of the corners and straight edge pieces of a jigsaw puzzle.

The foundation of Lean is flow logic and leveled production, yet it seems that few people or companies focus on these vital aspects of Lean. Why? In our experience, it's because these aspects are not really understood. But it is these very aspects of Lean that provide the stability required to achieve *sustainable continuous improvement*. We have found that eliminating waste is an *outcome* of implementing flow logic and leveled production. It is a natural result rather than the prime focus of attention. Like any jigsaw, one should start with the corner pieces followed by the straight edges—flow logic and leveled production—and only then put in the center pieces. These are the tools and techniques most people associate with "doing Lean."

In many organizations, Lean/RfS has provided a breakthrough in the company. Firstly, it removes much of the firefighting that is endemic in most companies. As a result, the organization gains a much-improved performance on parameters of efficiency, quality, and waste while at the same time acquiring lower inventories and higher customer service.

Secondly, Lean/RfS can change people's mindsets about operation strategies within companies. It can encourage greater teamwork and engagement of people, together with making more time available to work on improvements rather than constantly firefighting. Lean/RfS really does change people's behavior, and how they work with one another, for the better.

However, there have also been two common issues within companies after the organization has implemented Lean/RfS.

■ Firstly, people cannot see clearly how they can use the increased available time they now have, and what they can do to further improve flow—and hence performance—across the total supply chain.

■ Secondly, people cannot see how they can go on to apply Lean/R/S principles across every function to improve all business processes in order to effect a Lean transformation across the whole organization.

There is a risk that, having implemented Lean/R/S into the manufacturing process, a company will describe itself as "having done Lean/R/S." Unfortunately, Lean/R/S is *not* seen as the stable foundation on which far greater continuous improvement activities can be built in every area of the business—improving the quality and effectiveness of the service that each function delivers while reducing costs.

The objective is to have every function in the organization achieve a step change in performance—and they can, with Lean/R/S. These improvements need to be aligned, which is where policy deployment comes in, to ensure that the ultimate objectives are achieved: simultaneous improvements in market share, profit margins, employee motivation, and customer satisfaction. This is a tall order, but it is one that we will demonstrate can be done. Stopping corporate firefighting by moving from batch to flow logic in manufacturing is just the first step.

Kimberly-Clark (KC) has been working with Lean/R/S since 2005 and has witnessed some great results. They have also witnessed the issues we've already identified. To obtain the maximum gain possible, KC recognized the need to resolve these issues before they could integrate Lean/R/S into the whole of their business and total supply chain. The goal was to create a stable foundation for sustainable continuous improvement and enhanced business performance. This meant combining Lean/R/S with other improvement techniques and principles. The key aspects of this initiative were policy deployment and Lean leadership, and KC used different Lean experts in these areas to help them. KC has greatly benefited from applying Lean/R/S and integrating it into its overall business improvement process.

Ian Glenday (a coauthor of this book alongside Rick Sather) has continued to assist KC. He has also worked with many other organizations—in particular global branded FMCG (fast moving consumer goods) companies—in implementing Lean/R/S into their businesses. He found in these companies the same types of issues discovered by KC. This workbook is about how to apply and integrate Lean/R/S into large organizations so that it becomes part of the way the company does business, rather than an initiative, pilot, or flavor of the month in manufacturing. So, after the initial Lean/R/S implementation, when the firefighting has been greatly reduced and people find they really do have more time, it will help them achieve a companywide Lean transformation.

This workbook is *not* about the Lean/R/S journey at Kimberly-Clark; however, it does include many examples from their experience as well as examples from other companies that have implemented Lean/R/S. The objective of this workbook is to help people understand how Lean/R/S can become the foundation for achieving sustainable continuous improvement—to discover how they can use the *stability, repetitiveness,* and *structure* it creates as the basis for a total Lean transformation.

Ian Glenday and Rick Sather
December 2012

About the Authors

Ian Glenday started his Lean journey as a microbiologist running a plant producing enzymes from deep-culture fermentation of bacteria. It was here that Ian first began developing R*f*S concepts and principles for application in process industries.

After taking time out to gain an MBA from Bradford Business School in the UK, Ian joined the manufacturer Reckitt & Colman, where he led an MRPII project to Class A status in the company's pharmaceutical division. This experience offered Ian a valuable lesson in understanding why applying batch logic in MRP can cause problems.

Ian then moved to Reckitt & Colman's household and toiletries division, where he initiated and helped implement a pan-European supply chain strategy based on the Lean concept of "every product every cycle," before joining Colman's of Norwich as head of policy deployment, responsible for applying Lean/R*f*S thinking across the entire company.

Ian currently divides his time between working with Professor Dan Jones at the Lean Enterprise Academy, UK, where he is a senior fellow, and helping businesses around the world make their own Lean transformations through his company Repetitive *flexible* Supply Ltd.

Rick Sather is vice president, customer supply chain, for Kimberly-Clark Corporation's North America Consumer Products Division. In this role, he is responsible for service and efficient product flow from the end of manufacturing through the customer's retail shelf.

Originally from Wisconsin, Rick received a BS degree in industrial technology from the University of Wisconsin-Stout in 1985, and for the past twenty-seven years has worked in a wide range of supply chain roles. Rick's Lean journey began in 2005 when he first connected with Ian and began implementing Lean/R*f*S at Kimberly-Clark. Learning and applying Lean/R*f*S in direct-line roles has enabled Rick to establish a problem-solving culture focused on delivering exceptional outcomes for people, customers, and shareholders alike.

Chapter 1

Twenty-Five Years at Kimberly-Clark

Does This Sound Familiar?

My entire working career has been at Kimberly-Clark (KC). The company manufactures paper-based products such as tissues, toilet paper, and diapers with famous brand names like Kleenex®, Scott®, Andrex®, and Huggies®.

Throughout this time, I have held many roles in manufacturing, planning, and the supply chain. My current position is vice president, Customer Supply Chain North America. One of my early jobs at Kimberly-Clark, back in the mid 1980s, was production scheduler for Kleenex facial tissue. During my first week of training, the current scheduler told me that the first task to do each day was to check what had been produced the day before.

Very often, this would be a quantity that was different from what had been planned. As a result, he then changed the schedule for that particular day. It seemed very odd to me that this was how scheduling was done. I can remember the feeling I experienced each day knowing that the production people would not be happy about yet another plan change. This was my first experience of operating in a working world of short-term plan changes and firefighting. It was an experience that became the norm.

Of course, we tried to minimize the amount of plan changes, but these only seemed to scratch at the surface. A few months later, JIT (just in time) became all the rage, and we were being challenged to apply JIT techniques. I read the books, and they seemed logical—but there was no way that JIT was going to work in our environment. And it didn't. There were simply too many plan changes and too much short-term firefighting going on. It is laughable to think back, because JIT actually made things *worse*. We tried to bring materials "just in time" for production, but now, not only were we changing the schedule to react to issues with sales and production, we were also having to change the schedule due to shortages in materials.

Over the years, we pursued many other improvement initiatives besides JIT at KC, e.g., employee-involvement and quality-circle improvement programs. Both seemed like great ideas, but somehow they didn't quite work out as the books and experts said they should. "Just in time" became "just in case." Better to hold a bit more inventory than risk the stock outs that tended to happen when you tried to implement "just in time." People worked very hard at making improvements. Safety and quality improved. The rate of production increased. However, there always seemed to be lingering problems. Even when improvements were made, there were still big gaps between functions. People were still engaged in blaming each other when things went wrong, and they did go wrong. People seemed to be constantly caught up sorting out the issues caused by short-term plan changes. The focus was not on identifying the root causes for these changes and fixing them, but more on fixing the symptoms—known by everyone as "firefighting." The same issues seemed to keep cropping up time and again, despite everyone's best efforts.

Do You Face This?

Do you face conflicting objectives and targets? Do you face increasing pressure to grow sales plus reduce fixed costs and working capital while improving production and supply chain efficiencies in order to become a more responsive, flexible, and efficient supply chain? Meanwhile, the number and complexity of products continues to grow.

A Typical Example From KC

Diapers—"nappies" in the UK—are an important product category for KC. The last twenty years have seen enormous advances in the effectiveness of the product. Leak protection, form, fit, and function have all advanced dramatically to fit the stages of a baby's development. Today's diapers are a technological marvel with absorbent gels, fasteners, and leakproof leg seals. However, these improvements have also more than tripled the number of component parts in the product. In addition, the range of products has expanded. The number of sizes available has increased; features have been added to meet specific needs; and package configurations have grown to match the variety of retail alternatives. New product categories have been invented, such as Pull-Ups® training pants. There are even different colors for boys and girls. The combination of these factors has made the manufacturing process more complex. Manufacturing departments would like bigger batch sizes to give them longer runs. More product types mean more items to warehouse. Yet the overall inventory target has been reduced. Longer runs (to give higher production efficiencies) coupled with lower inventory (to reduce working capital) at the same time as delivering high order fill rates (to achieve excellent customer service) are in conflict with each other.

This is the conundrum that supply chain personnel face every day.

Have You Done This?

Have you spent millions of dollars on consultants and new IT systems, only to find that the benefits achieved have been questionable versus the costs involved?

We looked to new IT solutions to help enable us to deliver excellent customer service with high efficiency and lower inventory. In many cases, as complexity in the product range increased, we were just not able to deliver the targeted inventory reductions or other savings from these system changes. The computer program solutions available to address the issues of supply chain have grown enormously over the last twenty-five years. We now have more sophisticated systems, but that can also translate into greater complexity and a loss of user-friendliness. Despite having bigger, faster, and better-integrated systems, we still have the same old issues of sales forecasting errors, inventory inaccuracies, and bill-of-material mistakes. These are all issues that contribute to the short-term plan changes that lead to the firefighting we suffer from on a daily basis.

Have You Experienced This?

In a word, *initiatives*: the Theory of Constraints, MRPII (manufacturing resource planning), TQM (total quality management), Six Sigma, Agile, and many more. Driving for improvement has always been part of the culture at Kimberly-Clark. We have experienced a whole series of "initiatives" during my twenty-five years at KC. The champions of each all claim that "their" initiative is the one that will deliver a step change in performance. There are inevitably arguments between the advocates of different approaches on which are the best for improving performance. It seemed to me that there were certain similarities between each of these initiatives. Obviously they were all aimed at improving company performance, and they did achieve this, at least initially. At KC we applied each latest initiative with enthusiasm and commitment. However, not everyone in the company had the same levels of energy for implementing the "new" initiative. It tended to be driven by a few determined "disciples." When they moved on—either to another role or the next "new initiative"—the discipline of consistently applying the principles started to fade away. As a consequence, the gains made also slipped back. It was difficult to sustain any particular approach. It was much easier, and more fun, to embrace the next new initiative that came along.

I am sure my experiences are common to many people working in large organizations.

Searching for a Step Change

People talk about *wanting* a step change and how the latest initiative would deliver this, but my experience was that nothing ever delivered a *real* step change. Sure, KC improved—a lot. However, sustainable improvements tended to come from technological advances, better product designs and ranges, and increased machine capabilities. Actual sustainable improvements in the overall supply chain process itself were less forthcoming, despite great people working every day to make improvements. I was set the task of finding a step change in the way the supply chain was planned and managed.

At the time, I was working for the head of manufacturing in one of our divisions. He challenged me to search for a better way. It was not a mission to go out and find the next improvement program; it was a challenge to see if there was anything out there that could help us make a fundamental step change to the way we planned and managed the supply chain. It was a tall order and, as time went by, one I increasingly felt I could not achieve. There were lots of different views from consultants, academics, and people in other companies about what needed to be done. However, they all seemed to be just variations on the initiatives KC had already tried. They were also all aimed at *improving* the current process—not *changing* it. If you improve your forecasts, apply new software, increase inventory accuracy, etc., etc., everything will get better. Nobody questioned the fundamental logic that drove supply chain planning—economic order quantity (EOQ) or batch logic. This was the basis of practically all supply chain planning processes. It was the conventional wisdom that most people accepted as right. Most people—but not all.

One day, while browsing the Lean Enterprise Institute website, I came across a workshop titled "Lean in Process Industries" given by Ian Glenday. I had read several books on Lean, but virtually every example I read or heard about didn't seem to apply to our situation at KC. At that point, I certainly was not able to make the connection of how Lean could apply to Kimberly-Clark. Then I met Ian. He had a very different way of presenting things. He explained what was causing the persistent short-term plan changes and firefighting—why it happened in nearly all manufacturing companies and how EOQ or batch logic was *guaranteed* to change the plan every time it was calculated. It was a revelation. For the first time, I heard someone actually question the current logic: batch logic. Rather than focusing on *improving* how things were currently done, Ian wanted to *change* the way things were done. In all my searching, I had heard no one else put forward the reasons why batch logic is fundamentally flawed. The way it was described made perfect sense. It all seemed so, well, logical!

Having presented why the current logic caused short-term plan changes, Ian went on to cover the alternative: flow logic. He explained how "flow" was the foundation of the Toyota Production System. Yet I had always understood "Lean" as being primarily about eliminating waste. Ian referred to this as "chapter two." In other words, in reading books about Lean, chapter one was about "flow." However, people did not understand it or felt that what was described would be impossible in their business. So they went on to chapter two in the books. This was about eliminating waste and Lean tools like SMED, TPM, 5S, etc. Eliminating waste equals cost cutting, and people certainly understood what

Figure 1.1 Tissue machine in process of being changed over.

that was all about! Simple-to-explain tools to use on the shop floor to improve efficiencies—they understood these and could put them to immediate use. People had jumped to chapter two without thinking about how to apply chapter one. Yet chapter one was about creating the foundation to provide a firm and stable platform for all aspects of Lean.

It all made so much sense. I was excited to hear the practical details of how one went about changing from batch to flow logic.

The term used to describe the process of moving to flow logic was the *steps of leveling*. As I listened, I began to realize why people quickly skipped chapter one and went on to chapter two. The steps of leveling meant progressively making smaller and smaller batches with more frequent changeovers! This seemed ridiculous; it flew in the face of everything I knew about improving efficiencies. But Ian was insistent that shorter runs and more changeovers would make the supply chain and production more effective. This was very difficult to accept, especially for Kimberly-Clark. As a paper producer with big and expensive assets, KC has a long history of big batches with equipment being "run full." I asked Ian if he had ever seen a tissue machine—the first stage in making a lot of KC's products. The answer was "no," so I showed him a picture of a tissue machine in the process of being changed over (see Figure 1.1). "Fast changeover that!" I challenged Ian.

He was not fazed, although he did admit it might be a challenge. He explained that flow logic was not just about changing the algorithm the planning system was based on. It was much more than that. It would change the very way people behaved—both the way they worked together and the way they tackled issues. This really did start to stretch my credibility—it was too far-fetched a claim. How could a change in the planning logic alter people's behavior?

Driving back from the workshop, I kept thinking about what Ian had said about batch logic actually causing plan changes. It seemed so obvious, now that he had explained it. It made common sense. Yet the alternative logic of flow seemed ridiculous to apply in a process industry—shorter runs and more changeovers just did not make sense. The term *paradigm shift* kept coming to mind, the definition of which is:

> What today seems impossible to do but, if you could do it, it would fundamentally change what you do.

What we needed at KC was a paradigm shift in the way we planned and managed the supply chain: not *improve* what we currently do, but *change* what we do. It was the only truly different approach I had heard. It had to be worth giving it a try, and we did. The rest, as they say, is history.

Summary of Chapter 1

At KC we had tried numerous approaches, initiatives, technological advances, product innovations, and IT systems to improve our supply chain, our manufacturing, and company performance. And we had improved—a lot. Yet still we were plagued by short-term plan changes and firefighting. This was hardly a good platform for achieving sustainable continuous improvement and applying consistent standards in the way we did things. To make a step change to further drive performance improvements was going to mean doing things differently. The only really different approach I heard was Repetitive *flexible* Supply, or R*f*S, the term Ian used to describe changing the planning logic from batch to flow. Could this small but significant change lead to a transformation across the whole business? Would it actually impact people's behavior? Was it even possible to achieve this change in KC?

The answers are "yes," but with some pitfalls and lessons learned along the way. This workbook is *not* a history of R*f*S and Lean at Kimberly-Clark. It is about how one puts the various elements of Lean/R*f*S together. As we explained in the Introduction, using the humble jigsaw puzzle as an analogy, one needs to start with the corner pieces. Then one adds the straight edges before finally filling in the middle pieces to create the whole picture. The waste-elimination and tools aspects of Lean are all valid improvement techniques; however, they are, in our opinion, equivalent to the center pieces of a puzzle. Without the corners and straight-edge pieces, it is much more difficult to complete the whole picture. This workbook is about putting the pieces together in the right way, with examples from Kimberly-Clark and other companies.

Chapter 2

The Fundamentals of Lean/R*f*S

A Brief History

Lean/R*f*S—or Lean/Repetitive *flexible* Supply—is made up of a number of different components. Each of these will be explained in more detail in the sections following this brief history.

Originally, the term used was simply Repetitive *flexible* Supply, or R*f*S for short. Lean was added later for two reasons. Firstly, many of the components are the same or very similar to the Lean concepts and principles developed by Toyota. Secondly, most of the companies implementing R*f*S had had Lean initiatives running for some time. They wanted people to see R*f*S as a further development in their Lean program, not as just *another* new initiative. People started to use the term *Lean/R*f*S*, and, well, it just stuck.

It started back in the late 1970s when I was managing a biotechnology factory that produced enzymes. The plant ran twenty-four hours a day, seven days a week. It was a very complex operation with a high degree of variability, including types of products, fermentation times, yield rates, packing efficiencies, and sales volumes. It was consequently a nightmare to plan—and the plan kept changing. This created yet more work for everyone. Far too much time was taken up with replanning and then dealing with the consequences of the changes—more commonly known as firefighting. Not enough time was focused on root-cause resolution of issues and real sustainable improvements.

I reasoned that if we could develop a repeating pattern of production it would make things a lot easier. This pattern took some time to figure out, but the breakthrough came with the realization that the wisdom of long runs—based on economies of scale—exacerbated the problem of replanning. The reason for this? Because if sales for a product not currently being produced was higher than forecast, we had service issues. So the longer the production runs, the greater was the chance that sales would be different from forecasts.

This meant changing the plan to make that product. Cutting the length of production runs would help reduce the problem, and making everything more frequently would decrease the time between each production run for a particular product. This meant that the impact of having to change the plan due to sales being different to forecast was reduced. It just seemed sensible to make a plan that repeated weekly—to make every product every week in the same repetitive sequence.

It worked very well in the small biotechnology factory. Only later did I learn just how much this simple idea flew in the face of conventional production planning.

When I moved to a much larger FMCG (fast moving consumer goods) company in the early 1980s, I tried to explain the concepts of R/S to the senior management team. The idea seemed ridiculous to them. I decided that what was needed was a visual concept where they could physically see it for themselves. Supply-chain games to demonstrate ideas like Just-in-Time were all the fashion then. So, a game to demonstrate R/S was developed. It consisted of two teams; each team had the same make-up, including customers, sales director, planner, factory, and suppliers. The teams made little boxes to simulate the production process. Team 1 operated batch logic; Team 2, R/S logic. The game took about an hour and a half to run.

The first time we ran the game, my boss, Dr. Bill Walsh, invited a friend of his along to see it. His name was Michael Brimm. He was professor of organization and management at INSEAD, a business school in France. At the end of the game, Michael asked me what I thought R/S was about. "More frequent production gives lower stocks and higher customer service at the same time," was my reply. Michael said that he'd been watching the teams during the game and looking at how the contestants were reacting. In Team 1, he had observed people arguing, quite aggressively at times, and blaming each other for what was going wrong. The person playing the sales director kept fiddling the forecast data to manipulate what was happening in the rest of the team. Factory volumes fluctuated wildly. The people playing the suppliers did not have a clue as to what was happening. It was a scene of total chaos and a complete lack of efficiency.

Team 2, however, was completely different. People were cooperating. They had moved their tables and chairs closer together to make things easier. They were helping each other. The suppliers had agreed with production to deliver directly to the factory rather than to the material warehouse. The scene was one of calm and effectiveness. Michael observed that R/S seemed to affect people's behavior in a positive way. Every time we ran the game, the same thing happened. The question was, would this behavior be repeated in a real working environment? If it was, then the potential of R/S was far bigger than just a different way of planning.

The good news? When R/S was implemented for real…it *did* happen.

Because of what they witnessed during the game, senior management in the company backed the implementation of R/S across Europe.

As the rollout progressed, three new aspects of R*f*S started to become visible for the first time. We were analyzing the product sales in each country using the Pareto principle (also known as the 80–20 rule), where 80% of the sales typically came from 20% of the product range. It was this 20% of the product range that we were focusing on for implementation rollout. As each country's sales were analyzed, another, much more significant set of figures kept reoccurring. Half of the sales came from just 6% of the product range, plus an amazing 95% of sales came from just 50% of the products.

Every time the analysis was done, the same results occurred. The figures were much more significant than the Pareto analysis. It meant that just 6% of the products needed to be put into the weekly repeating pattern of production and distribution in order to have half the sales volumes in an R*f*S flow. This was a far smaller task to achieve in the rollout than the 20% of products we had been considering using the Pareto principle. Working with a close colleague and friend, Carla Geddes, we developed a "language" of colors for the different categories of products based on their cumulative volumes. We also decided that a name for the analysis was required. Pareto analysis was named after the person who recognized the mathematical principle of 80/20, so Carla suggested we call this new analytical tool the Glenday Sieve.

The second aspect that came to light as we continued to focus the rollout on the top 6% of products—or "greens" as we called them in our new language—was their demand variability. Or rather, the lack of it. The general perception in the business was that sales had high variability and low predictability. Sales forecasts were a joke, especially for the biggest-selling green items. Sales, when expressed in volumes, seemed to vary quite a lot. However, when one looked at *percent* variability, it was actually quite low, often no more than ±20% to 30%. So the perception was a myth. The sales of the greens were relatively stable, meaning that changes to the weekly production schedule of greens due to sales fluctuations were even less likely to happen.

The third aspect that started to emerge was a big surprise. After initial skepticism from most people, and operators in particular, people began to say that they really liked the R*f*S approach. The repeating fixed weekly pattern created routines—and people like routines. Not only did they like it, but performance also kept improving. Each week they found little ways to make things run smoother with less effort and greater effectiveness. It was a phenomenon no one expected. Batch logic focuses on machinery. It aims to improve by creating what is known as "economies of scale." Long runs with fewer changeovers make the machines more efficient. Repeating weekly plans created routines that gave people a structure, pattern, and stability, all things that they liked and, coincidentally, also helped to improve performance. This new phenomenon focused on the people, not the machines. We called it "economies of repetition."

I was fortunate to visit Japan in 1990. It was a study tour of Japanese manufacturing led by Yoshiki Iwata from Toyota Gosei, one of the first suppliers to be taught the Toyota Production System (TPS). One night we were together in a bar in Tokyo. To make conversation, I explained about R*f*S and how we were running every product every week in a fixed weekly schedule. Iwata asked me to explain again, to be sure he had understood. That is when I first heard the term *heijunka*. Iwata explained that that was exactly how Toyota had started—with fixed cycles, the same production sequence repeated. He explained that the initial aim was to create stability, and that flexibility and responsiveness to actual demand came later. He described to me a five-step process known as the Steps of Leveling—the foundation of TPS. I was already at step two through intuition, coincidence, and sheer luck. I had not recognized or realized that there were other steps in the process. It was a revelation to me when Iwata described all five steps.

The next few years were spent implementing R*f*S, with various degrees of success, in the companies that employed me. Since 2001 and working as an independent, I've been trying to encourage more people in companies to understand the concept of Lean/R*f*S. Now, more and more people are using it to help them improve performance in their companies.

Lean/R*f*S is a very different approach than the conventional wisdom of supply chain planning and management. We have had to experiment, to "try by doing," and not always successfully. It's been a struggle to get the whole company engaged, for example: getting the finance department to understand how R*f*S could lead to a different and better way of product costing; getting the purchasing people to appreciate how R*f*S could lead to a truly cooperative and mutually beneficial relationship with suppliers; and last, but by no means least, getting marketing and sales to see how they could use this to benefit customers and consumers—as well as increase market share and margins—in short, to *beat their competitors*.

This Lean/R*f*S workbook aims to bring to you these lessons we've learned and help you engage *all* the functions in your business. We also want to show how Lean/R*f*S is more than just a different way of planning: *It's a way to help transform your business into a truly Lean enterprise.*

The Key Components of Lean/R*f*S

Batch Logic Issue

Batch logic, or economic order quantity (EOQ), is the basis of most planning and ordering systems. Yet this logic is fundamentally flawed for two reasons. Firstly, it is responsible for causing the firefighting that is endemic in most manufacturing companies. Firefighting becomes inevitable when plans are changed after they have been issued. We will show why plan changes are *guaranteed* when using EOQ logic. Secondly, it also causes unnecessary *variability* in what is planned and executed.

Batch logic uses a target stock figure to calculate what is required to be produced or delivered. If the underlying data have changed since the last plan was calculated, then one gets a different result. Let's take a look at the underlying data. How often do actual sales differ from what was forecast? How often does production make a bit more or less than was planned? How often do you come across inventory inaccuracies in the warehouse? When we look at the real world, the chances of the underlying data remaining unchanged from one plan calculation to the next are zero. So, each new ordering or production plan calculation generates a different answer than the previous one. Certainly a lot of effort is made in companies to get their data accurate. For batch logic *not* to calculate a different plan every time would require all of the underlying data to be 100% accurate and consistent.

This is NEVER going to happen.

Sales will be different than the forecast; production will make more or less than planned; inventory inaccuracies will occur. What is required is a logic that can cope with a *degree* of data inaccuracy.

Batch logic also creates unnecessary variability. The reason for this is that the initial calculation of the requirements is usually rounded up to create an "economic order quantity" (EOQ). Examples include pressure to fill a truck, to work to a minimum run length, or to produce a volume to trigger a sales discount. These pressures introduce artificially higher demand, which then translates into artificially lower demand at a later stage in order to bring the stock back to the target level. The result is peaks and troughs in demand, otherwise known as the Bullwhip Effect.

The following example illustrates just how bad this problem is. Figure 2.1 shows actual consumer sales of a single product passing through the tills of a major retailer over the fifty-two weeks of a year.

Apart from a peak at Christmas and a dip in January, the rest of the year is practically flat. Underlying demand is surprisingly stable. Knowing this, we would expect the retailer's planning systems to show stable orders to its supplier. But the actual orders received by the manufacturer are shown in Figure 2.2.

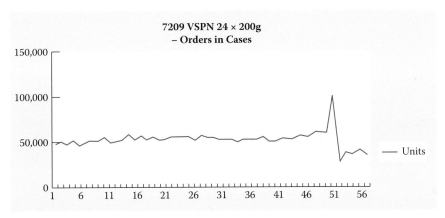

Figure 2.1   Actual consumer sales.

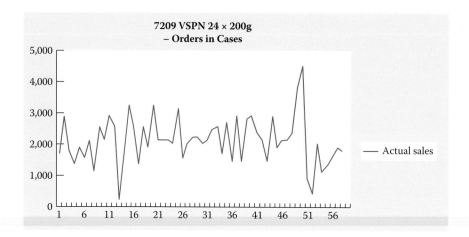

Figure 2.2 Actual orders received from the retailer.

One step in the process and EOQ converts a level demand into orders varying between 300 and 3,000 cases!

It is indeed strange that most ordering and planning systems are based on a logic that has not one, but two fundamental flaws. That so much effort in companies is directed at trying to achieve 100% data accuracy—which is essential for batch logic to work—yet it's totally impossible to ever achieve this.

Is it possible that the conventional wisdom of planning using EOQ batch logic is fundamentally wrong? Surely people involved in designing and operating these planning systems would have recognized this if it were true. They could not have all got it wrong—could they?

Conventional wisdom can be wrong.

It's surprising just how wrong conventional wisdom can sometimes be; how large numbers of people whose experience and skills one would have thought made them well qualified to spot when conventional wisdom is wrong, yet they still manage to miss it. A good example of this concerns stomach ulcers. Medical consensus was that stomach ulcers were caused by excess acid, stress, and poor diet. The cure was to remove the ulcer surgically. Often the ulcer would reoccur in the patient. Two microbiologists, Robin Warren and Barry Marshall, had a different opinion. They believed the condition was caused by a bacterial infection in the stomach.

Their view met with ridicule from doctors because it flew in the face of many years of accepted medical knowledge. It was argued that the causes of ulcers were well understood. Besides, the stomach was too acidic for bacteria to survive. No one would believe them. So Marshall decided that a radical experiment was required. He drank the bacteria in a suspension and gave himself an ulcer. He promptly cured it with a course of antibiotics, a cure that would only be effective if the bacteria were indeed the culprits. The name of the bacteria is *Helicobacter pylori*. What was seen by doctors the world over as a heresy had been proven to be right. Conventional medical wisdom was wrong. In 2005,

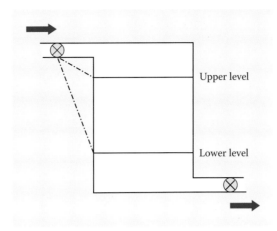

Figure 2.3   Buffer tank.

Marshall and Warren received the ultimate badge of acceptance—the Nobel Prize for Medicine.[*]

Alternative Logic of Flow

There is an alternative, and its name is "flow logic." The warehousing of stock is intended to balance the differences between supply and demand. To work effectively, warehouses need to operate in the same way as the real buffer tanks one finds in production, as shown in Figure 2.3.

It is called "flow logic" because the liquid flows continuously through the tank. When there is a difference between the input and output flow, the level in the tank goes up or down to compensate for this difference. If the level exceeds the upper or lower control limits, there is an adjustment of the input flow to bring the level back within the limits. Buffer tanks are fully automatic; no one needs to do anything once they are set up. They control differences between the input and output. Warehousing should—but usually doesn't—work in the same way. The inventory level in the warehouse should be allowed to fluctuate between limits without any interference. Items should continually flow into the warehouse.

This seems ridiculous to most people. How can there be a continuous flow? Items are produced and delivered in batches. Most warehouses operate batch logic with a fixed target stock level. Even when there are min/max limits, there is no continuous flow. Items are delivered into the warehouse in batches.

It would seem impossible to replicate buffer tank flow logic in warehouses. Yet that's exactly what Lean/RfS aims to achieve. It flies in the face of conventional supply chain logic.

---

[*] One of many articles naming Robin Warren and Barry Marshall as discoverers of *Helicobacter pylori* as a cause of stomach ulcers and winners of Nobel Prize for medicine in 2005: *The Lancet, 366*(9495), 1429, 22 October 2005. doi: 10.1016/S0140-6736(05)67587-3

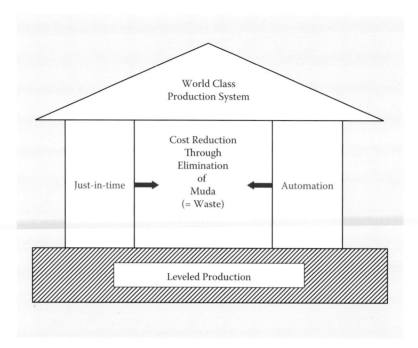

Figure 2.4   The Toyota Production System house.

Lean and Leveled Production

When I visited Japan and met Yoshiki Iwata, he showed me the TPS "house" as originally drawn by Fujio Cho of Toyota in 1973 (see Figure 2.4). Leveled Production forms the foundation. As Iwata described it, leveled production is an evolution of steps. The initial steps are designed to create stability, which greatly supports achieving *sustained continuous improvement*, leading to better performance. Once better performance is realized, it becomes possible to invest more effort to increase flexibility and responsiveness to demand. This sequence of steps is vital. You have to create stability first. Then you can progressively match output much more closely to actual demand.

This diagram highlights another important point. To many people, the prime focus of Lean is waste elimination. Yet eliminating waste is in the middle of the diagram. Why do many people and organizations focus on the waste-elimination aspect of Lean while ignoring leveled production? It would appear that there's a lack of understanding of how Toyota progressively developed and implemented the steps of leveling. The term *Leveled Production* implies the same amount all the time, but how can that be achieved when demand is not stable? On the other hand, people easily understand waste elimination. It's something they have been doing for years. Now, value-stream maps show them lots of other sources of waste to work on. They see this along with the tools and techniques as the prime focus of Lean.

The aim of the initial three steps of leveling is to produce products in a fixed schedule that is rigorously flowed. Same products, same volume, on the same equipment, at the same time, and in the same sequence every cycle—step

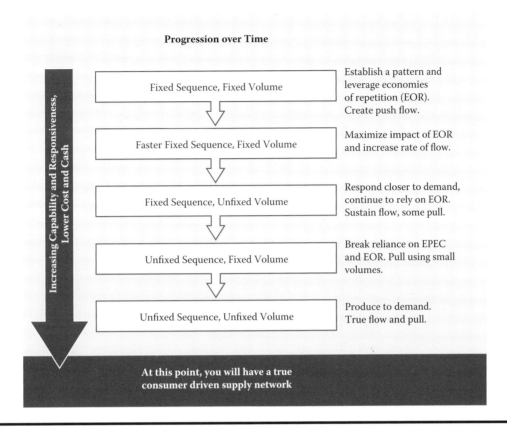

Figure 2.5 Progression over time.

one in a twice-a-month cycle, step two in a weekly cycle, and step three in daily cycles.

This seems impossible and ridiculous to most people. It's also counterintuitive.

The key objective of these first three steps is to create stability. Stability is the foundation that makes it easier to carry out continuous improvement and root-cause problem solving. These improvements are then invested in achieving faster and faster cycles. Progressively having faster cycles also means production can get closer and closer to actual demand (see Figure 2.5). Steps four and five focus on producing exactly to actual demand; firstly, in very small fixed quantities, then, just one item at a time—known as "one piece flow." This is the ultimate objective of Leveled Production.

Economies of Repetition

In steps one to three, by repeating the same pattern of *every product every cycle* (EPEC), we generate a phenomenon we have called Economies of Repetition (EOR). EOR works by building on two basic traits of human behavior. The first is the learning curve. When people do the same task repeatedly, they naturally get better at it. This isn't about applying improvement tools. It just happens.

The second trait is that most people like routine. Some people see this as being the same as learning curves, but our desire for routine is something different. To witness this for yourself, go into your organization's restaurant at lunchtime. Draw a map of where everyone is sitting. Now go in the following day at the same time and draw the map again. You will get practically the same map, with most people sitting at exactly the same tables, mostly even in the same seats! People live their lives by routine. They like the stability, security, and predictability that a routine gives. They often find it stressful if their routines are disturbed. That's one reason why short-term plan changes demotivate people. If you give people routine, they will feel less stressed, more relaxed, and better motivated. Also, they'll find it quite natural to keep making little improvements to make their work more effective and efficient. And little improvements being made by everyone across the company add up. The really good news is that experience shows the improvements do not plateau—they just keep getting better and better from the increased learning curve coupled with routines.

There's a third reason why "every product every cycle" is so important. It creates an environment where standardization can flourish, and without standardization you cannot generate sustainable improvement. It is very difficult to achieve standard work when short-term changes are frequent and every week the plan is different, which is what happens with batch logic. In a firefighting environment, the operators' way of coping is for each to have their own way of doing things. It's much easier to accept the need for standard work in a stable environment of fixed-sequence cycles. Standard work starts to come naturally because when people do the same things in the same sequences, they start to agree on the best way of working. When every day is different, this is far more difficult.

The benefits of economies of repetition are surprisingly rich.

Glenday Sieve

A useful analysis tool we now refer to as the Glenday Sieve separates products (SKUs) into groups based on sales volume (or value if this is more appropriate). The results shown in Table 2.1 are typical. Many people insist that it is impossible for 6% of the product range to account for 50% of the sales volume in their business. Yet when the analysis is done, it is invariably found to be the case. These

Table 2.1

Cumulative % of Volume	Cumulative % of SKUs	Color code
50%	6%	Green
95%	50%	Yellow
No number linked to Blues		Blue opportunities
Last 1%	30%	Red

results are intuitive for some and genuinely shocking for others. Either way, the power is in the indisputable information it provides.

It's not too difficult to develop a fixed-sequenced cycle for just 6% of the products. One can then value-stream map these "green" items to help unravel the "spaghetti" pathways one usually finds through the supply chain. The result is a "green stream" for these few high-volume products, with shorter through-put times, better material flows, fewer activities due to routines, and much lower inventories possible for the green SKUs and associated materials.

The "yellow" SKUs illustrate where to direct your capability improvement efforts—changeover reduction exercises and smaller batch sizes to make it easier to introduce these products into the fixed green cycle. This typically results in a staggering 95% of the total volume yet only 50% of the product range now running in a fixed-sequence "every product every cycle" schedule.

There is no number associated with the "blues." These are not SKUs, but they do show the complexity issues that exist; these are the factors consumers either do not recognize or do not care about, as they don't add value, only cost. What opportunities are there here for the harmonization of raw materials and packaging so that the final product appears different to customers but makes these SKUs easier to include in the cycle? For instance, one could have differently colored bottles but with the same shape, thus making changeovers easier and faster. One could introduce labeled, rather than printed, cartons to reduce the number of different cartons one needs to plan and hold inventory for. These may sound like simple things but they can make big differences in getting these products into the cycle, making them flow, and reducing overall costs without impacting what the consumer values in the product.

The "red" SKUs will need to be carefully reviewed to determine their real cost to the business. The impact of these products to the total supply chain and overall company costs, including overheads, must be understood in order to be certain that the benefits of selling them genuinely outweigh the costs. It's not uncommon for companies to recognize through this analysis that they actually have two distinct businesses. One is high-volume products—the greens—for which the plant, business processes, and performance measurement systems were designed. The other is a low-volume operation—the reds—with often more complex products to produce. The same equipment, processes, systems, and KPIs (key performance indicators) are being used for both "green" and "red" SKUs, to the detriment of customer responsiveness, efficiencies, inventory levels, and profit margins. Is it really a sensible business practice to continue running small-volume red items on equipment that was originally designed for high-volume green SKUs?

Central Limit Theory and Buffer Tanks

An understanding of the Central Limit Theory is also helpful in creating leveled production and flow. Put simply, bigger sellers usually have less

Table 2.2

Week	Product A			Product B		
	Actual sales	Volume variation	Percent variation	Actual sales	Volume variation	Percent variation
1	15745	472	103%	345	70	126%
2	15785	512	103%	256	−19	93%
3	14230	−1043	93%	410	135	149%
4	14896	−377	97%	137	−138	50%
5	16005	732	105%	335	60	122%
6	14975	−298	98%	167	−108	61%
Average	15273			275		

percentage variability than smaller sellers. This is because, in general, the sales of the bigger volume items are coming from a larger number of customers, so the variability in demand from individual customers is balanced out at the aggregate level. Correspondingly, smaller sellers come from fewer customers, so they don't have as much of this balancing out, resulting in higher *percentage* variability.

Ironically, because people generally measure production and sales at a unit level, not at a percentage level, many consider the bigger sellers to have the greatest variability. This is because the volume changes are large. The volume differences for the smaller sellers appear quite tiny. However, when one looks at percent variability, one gets a very different view of the variability that is occurring. The figures shown in Table 2.2 are for actual sales over a six-week period for a well-known brand of yogurt, comparing two products with very different sales volumes. This is more easily seen if one converts the data into a graph (see Figure 2.6).

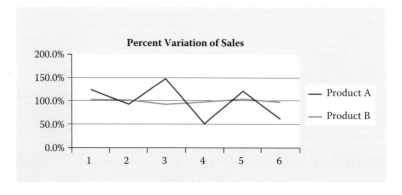

Figure 2.6   Graph of percent sales variation vs. averages.

How big a buffer tank is needed to cover the level of variability in the green product? The answer is, not very big at all. In fact, plus or minus half a week around average sales is more than sufficient in most companies to cover their sales variability for green SKUs. With the buffer tank in place to absorb demand variability, production will be stable. It makes sense to then produce these green products in the same amount every cycle at the same time in the same sequence, based on the average sales forecast. Typically, one takes the sales forecast over eight weeks to calculate the average.

The buffer tank will not cope with every situation. Sometimes one will need to change the plan, but it will be substantially less than when using batch logic.

Rockbusting

The first objective of R/S is to create a stable, fixed "green stream." This provides a "drumbeat," a rhythm that can be followed up and down the supply chain, so that materials, information, and money can "flow" smoothly, reliably, and quickly. Rocks in the green stream are anything that disturbs this flow. At its simplest, Rockbusting comprises two elements. Firstly, it is about identifying the rocks and then applying problem-solving techniques to help resolve the root causes. Secondly, it is about focusing on achieving conformance to plan, to produce *exactly* what was planned at the outset. These two elements are aimed at protecting the green stream to ensure that the flow gets better and better, and more reliable, with less inventory and non-value-adding activities. Having a repetitive pattern makes identifying the rocks and their root cause much easier to achieve than an environment where every week the plan is different, and then usually changes. It also helps in achieving greater conformance to a plan when the plan is the same in every cycle.

Rockbusting—named by an operator as a bit of fun after the hit movie *Ghostbusters*—is just a term to make sustainable continuous improvement a little more engaging for people. The underlying tools and techniques are the same as in any continuous-improvement initiative. It is the weekly repeating pattern—rather than different plans each week that are liable to change—that makes root-cause resolution and higher conformance to plan easier to achieve.

Summary of Chapter 2

This chapter of the workbook has been a quick explanation of the various components of Lean/R/S, in order to demonstrate that it is a mixture of existing ideas and concepts brought together to form a cohesive approach.

This approach has three key aims

1. To stop the firefighting that is endemic in organizations where planning is based on batch logic

2. To demonstrate how to achieve stability and flow as a foundation for sustainable continuous improvement
3. To apply these ideas and concepts throughout the entire organization to enable them to make a transformation into a truly Lean enterprise

We have been successful in helping companies implement the first two aims. The third has seen some success with different lessons learned within different companies. This workbook aims to bring these learnings together.

Lean/RfS is by no means a total application of Lean thinking, concepts, tools, and techniques. What it does more than anything else is help people in companies recognize the root cause of their firefighting—batch (EOQ) logic—and provide guidance on how to create stability, which is the very foundation of the Toyota Production System.

The next chapter describes what a Lean transformation into a Lean enterprise could look like when the firefighting is eliminated—*not just in manufacturing, but also throughout the company.*

Chapter 3

How It Can Be

Creating a step change in an organization is an exciting but daunting thought. People normally work in only a relatively small area of the business, and actions tend to be limited to that area. To make a change that is going to impact not just your own area, but across all the business—one that actually changes people's behavior—is challenging.

After the initial workshops at Kimberly-Clark to implement Lean/R∫S, Jayne Kelly, who was at the time KC's planning director, was interviewed for a business journal article. This is her comment about the experience of running with Lean/R∫S.

> The energy that it creates is amazing. Realizing far greater potential than any of us previously thought possible is very exciting. Imagine our company if we do all the things identified! But it is "scary" as well…in the sense of WOW! Are we ready to do all those things that we now know we can do?

To give you, the reader, a picture of what implementing Lean/R∫S can mean to a business, the following section presents a series of examples and anecdotes. They are not in any particular order and are not intended to be a "route map" of what to do. They're just some examples of what can be achieved. Implementing Lean/R∫S creates a paradigm shift, and the impact this has can be quite dramatic—on people's behavior as well as on results.

Some Examples and Anecdotes

Kimberly-Clark ran a 5-day workshop at a facial tissue manufacturing plant in Huntsville, Canada. The plant is the only one in North America that produces a smaller sheet size of facial tissue called Kleenex® Junior® facial tissue. It requires a significant changeover to and from the smaller sheet size, as almost all processes are involved and impacted, including the blades that cut the tissue to the right length, the carton maker, and the case packer. Prior to the workshop, the required changeover took five hours.

The Junior facial tissues were produced in a campaign-type cycle every eight weeks. Everybody hated that week. Not only was the changeover long, but the ramp-up to work out the start-up issues also took days to achieve normal speed and efficiency. The warehouse at the Huntsville plant is fairly small, so the large batch of Junior facial tissue produced every eight weeks was stored at a rented overflow warehouse. From there it was shipped to KC Distribution Centers. No one ever questioned why we needed the rented warehouse or the cost of extra handlings and shipments it caused.

One of the targets set in the workshop was to reduce grade-change times. We did not set a target to specifically establish a weekly production cycle. However, the green-stream team went to the mill manager, telling him they wanted to produce Junior facial tissue every week. This caused what we call a "side bar," a small group meeting in the mill manager's office to decide whether or not to attempt a major breakthrough, beyond previous expectations of what could be achieved. They don't happen all the time, but in some workshops they are necessary to work through people's natural skepticism and concerns. In this case, it was hard for the mill manager to agree that we should come out of the workshop producing Junior facial tissue every week. "Why can't we try it every four weeks or every two weeks? Our efficiencies will drop too much if we have to do it every week," was the protest. The green-stream team had calculated that they could meet demand on a weekly cycle if the grade-change time was reduced by 50%. After much debate and argument, the decision was made that we would go for it. Getting this alignment was (and is) very important, as the workshop teams and site leaders must be aligned to such a huge shift in the operation.

Coming out of the workshop, this was the first time within Kimberly-Clark where an entire plant was operating on a weekly cycle. It remains one of our best examples of where traditional thinking of long runs and large batches was changed to Lean/RfS logic of small batches and repetitive cycles. This led to the creation of *economies of repetition* (EOR) as a phenomenon that drives improvement.

The following is an e-mail received from the Huntsville Operations Manager after Lean/RfS was implemented at his factory. We think it pretty much sums up the multiple positive impacts that Lean/RfS can have in manufacturing.

> Hi Rick,
>
> RfS has been a tremendous success for the Huntsville site. Since the RfS start-up, productivity has increased 20%. The steady upward trend in rate of operation is a testament of the power of a frequent, stable, predictable repeating schedule. I am a believer and would fight any efforts to go back to the prior scheduling chaos!
>
> We continue to run the one-week cycle including Junior® facial tissue starting every Wednesday morning at 8:00 a.m. Grade-change time into

Junior® from end of last grade to first cartons through the casepacker averages approximately two hours. There have been weeks with the grade change taking less than eighty minutes. When the grade change time stabilized at less than two hours, the focus changed to rapid start-up with low waste. Most weeks, production ramped up to full rate within a couple hours after start-up.

We've been digging into the productivity improvement "toolbox" using standard techniques fitted into the R/S structure. SMED, TPM, and equipment improvements are all used to improve rates of operation. The approach has been practical, with operations and technical team members' involvement. In particular, the line mechanics and electrical support are fully engaged with the operators in making improvements regularly. Efforts have concentrated on eliminating "rocks," reducing equipment variables, and stabilizing the process.

A major shift in thinking is the idea of "making the cycle" versus daily rates of operation. Instead of measuring rates of operation, we measure where we are in the cycle—ahead or behind by hours. As for challenges, R/S continues to show us weaknesses in equipment, proce-dures, and people. Example: the facial tissue grade 280s continues to be a challenge. Given it is run every week, we know operating issues are independent of operating team, materials, current procedures, and equipment set-up. Plans have been developed to methodically address observable barriers to flow to improve rates of operation. Although 280s rates of operation are stable and predictable, they are low compared to other grades.

As I stated at the start, the rates of operation continue a steady upward positive trend. The stable, predictable, repeating schedule contributes to problem solving and solution implementation. I am an R/S believer.

Jeff

There are some points to bring out here. Firstly, there is the shift from mea-suring rates of operation—that is, efficiency—to making the cycle. A key aspect of Lean/R/S is to *stop* focusing on the line efficiencies. Production's prime focus should be on meeting the plan. The aim is to remove the variability of actual production versus plan, one cause of plan changes. When production focuses their efforts on making the plan, we *guarantee* that improvements in line effi-ciency will follow.

Next is how the stable, repeating plan has facilitated both the use of stan-dard improvement techniques and the increased ability to identify and resolve issues. Both of these will result in further increases in output rates. It is no accident that Toyota saw leveled production as the foundation; it provides the platform for other Lean tools and root-cause problem solving to be far more effective and sustainable. That's something that is very difficult to achieve in a batch logic environment, where the plan is different every week *and then changes frequently.*

Lastly, this demonstrates the benefits in *manufacturing.* This is just the beginning. Having achieved a step change in performance in manufacturing, one should be asking two questions. Firstly, *"How can this improvement best be used to beat the competition?"* Lean/RfS is not just about "getting better." It is how one uses the improved performance to increase competitive advantage, market share, and profit margin—to grow the business. Secondly, *"How can one apply the principles of Lean/RfS to all business processes?"* How could "green stream" concepts be used in other functions to improve their processes—not just to have Lean/RfS in manufacturing, but also to achieve a *total business Lean transformation?*

Impact on Behavior

The effect that a fixed routine can have on people's behavior is quite amazing. Here are comments made by Steve Ackroyd, Lean/Six Sigma improvement manager at 3M.

> Fundamentally, people want to do a good job. But if operators' routines are repeatedly changed, they lose interest. And that's what constant rescheduling does all the time: People do not like short-term plan changes. With Lean/RfS and flow, it is a much better life due to less frustration. Nowadays people spend less time arguing with each other and more time solving core issues. Informal relationships have improved. People simply like working with each other more.

The scale of the negative aspect of short-term plan changes on people's behavior should not be underestimated. People like routines. This was brought out in somewhat graphic language in an e-mail from a site manager at a KC plant in Australia.

> G'day Ian,
>
> I am sure "easy" was *not* a word that sprang to mind when we first started fixed cycles. There was a general view that I was insane, but having said that, the crew would assassinate anyone who now tried to take their fixed cycle away. I'm probably a certifiable "early adopter"—the things you say seem to make so much sense.

At one of the early RfS 5-day workshops in KC, we encountered an opportunity that greatly reinforced the benefits of flow as well as how it can impact thinking and behavior. What is most amazing about this story is that what now seems a very obvious thing to do was, at the time, not obvious at all. In fact, the idea flew in the face of conventional KC wisdom.

In the production of tissue products, there are two main manufacturing processes. The first is a large tissue machine that takes fiber, chemicals, and water to form the base tissue. This comes off the machine as a large "parent roll" that is about twelve feet wide and six feet high. The parent roll is then cut up and packaged on a converting line into consumer-useable products such as toilet paper, paper towels, and facial tissue.

Since the tissue machine assets are large and expensive, the conventional wisdom was to operate them at a high level of utilization. In addition, they can be difficult to change over from one grade to another. So, long production runs were the norm in order to minimize changeovers.

During the early part of the 5-day R/S workshop, some of the operators said that older parent rolls (>24 hours) did not convert as well as "fresh" parent rolls. After they aged, the tissue properties changed enough that converting efficiency decreased and waste increased. Prior to R/S, our production planning processes scheduled both sets of assets separately. Parent roll inventory was an input into the converting schedule. It was not uncommon for there to be disconnects between the planning of the two operations. This resulted in swings in the inventory of the parent rolls, leading to some parent rolls being days old when used in converting.

As we set the targets for the workshop, this insight from the operators was very useful. We used it to set a target for parent rolls so that they would be no more than twenty-four hours old when used in converting. We did not tell the team that they needed to change over the tissue machine every day, meaning they would need to drastically cut the grade change time. We just set the *outcome target* that was needed and let the team themselves determine how to achieve it. It is important to set a target of what is *required*, rather than how to *achieve* it. Even worse is to set a target based on what people *believe* can be achieved, given the current thinking at that time.

When running a 5-day rapid implementation workshop, it's always interesting to witness what happens when teams are agreed and then start working together. Some team members think we are totally out of our minds for setting what appears to them to be impossible targets for them to achieve in that week. Others can't wait to get going. As usual, the team working on this was a mix of people who normally would not work closely together for a whole week. It included operators who ran the tissue machine, maintenance associates, process engineers, and management. They obviously knew that to operate with less than twenty-four hours of inventory would mean they needed to improve the changeover time on the tissue machine. What they also realized was that they needed to develop a way to trigger the changeover so that they would not run out of the right grades of parent rolls for the converting operation.

Through the course of the week, the team experimented with several changeovers and was able to reduce the time of the grade change from over two hours to just fifteen minutes—way beyond what people had previously thought possible. Previous attempts at reducing the change times had not set an output target. Nor had they put together a mixed team for a whole week. This meant that the tissue machine could support the converting operations with fresh parent rolls of each grade required every day. As the team ran the tissue machine to their new way of operating, they discovered something that was a big shock. The triggers from converting consumption did not support continuous running of the tissue machine. In other words, the tissue machine did not need to run full to meet demand—it could be switched off some of the time. This became quite a topic of discussion, indeed of concern, during the workshop as, traditionally in KC, tissue machines were *always* run to the maximum because they are such a big, expensive asset. Ian Glenday ended up spending a fair amount of his time with the team and the site manager on getting their agreement to stopping the tissue machine when the parent roll stock hit twenty-four hours of inventory. Stopping the tissue machine was a big step for both the mill team and KC. It had never been done before.

But they did it, and the result was a big success. With "fresh" parent rolls, the converting lines ran much better, meaning higher throughput and lower waste. Now, due to this improved performance in converting, more parent rolls were required. That meant running the tissue machine more. This gave extra production that sales could sell, leading to more profit for KC. Operators had an easier time running the tissue machines and converter lines, the company created more sales from the same assets, and the KPIs (key performance indicators) in production improved—a win-win all round. Breaking the long-term KC conventional wisdom of running the tissue machine full was a paradigm shift in both thinking and behavior. This change would never have happened without the implementation of Lean/R*f*S and the challenge to existing paradigms that it creates.

"How It Can Be" in an Office Environment

Lean/R*f*S is not just about manufacturing. Here's an example from KC of "how it can be" in an office environment.

After a major IT system update, the export orders team found themselves in a difficult situation. Instead of making the process easier, the new system had many issues that still needed to be resolved. So workload for the team increased, and frustration in the team was high. They were working weekends, with people working sixty–seventy hours a week, *every week*. The issues from the major system upgrade were known about, and there was a lot of stabilization work underway, but progress was slow. It was decided that even in the midst of this tough environment we needed to conduct a workshop to help address the issues.

We ran the workshop. The team developed their future state map of their processes with a Kaizen plan to move from the current to the future state. The team decided that one of their KPIs to measure implementation success was the number of hours people were working each week. Their key objective? To reduce the hours they worked back to an acceptable level.

At first, progress was slow and it seemed like the team would never be able to step forward. Then, it started to "happen." Small improvements through repetitive routines linked to problem solving started to reduce errors and rework along with the wasted time and effort it caused. The team had developed a visual management process to track their plans and a simple chart to track hours worked per week by each person. At the beginning, this tracker was "red" for the entire team. The team started seeing some major improvements. More and more team members were green, that is, within target of their work hours. The team had turned from an overburdened, highly frustrated group to one of the most motivated, high-performance teams I've seen. They meet every Wednesday at 2:30 p.m. in front of their planning board for less than thirty minutes. The team owns the plan; they have no computer record of it—it's just how they now "do" their work.

The reason for wanting to share this story is to make two points: Firstly, Lean/R∫S is *not just about manufacturing*. The principles can apply to *any* process. Secondly, *it is really all about the people*. It's about creating the environment where the green routine activities are separated from the variable red ones to greatly reduce firefighting and the chaos this causes, so that people have a greater sense of control and can make a difference. *It's simply a better way to work.*

Impact on Problem Solving

To make a fixed repetitive cycle work, problems have to be solved when they arise. Issues don't just go away with Lean/R∫S. This can mean big changes in the approach taken. Schedule breaks—changes to the green stream—must be highlighted and the reasons investigated. The following are comments made by Jayne Kelly, at the time KC's planning director:

> Prior to Lean/R∫S we would switch over to another product—change the schedule—if we ran into difficulties. This would be done by the team leader, and the mill manager more than likely would never even hear about it. It was the norm. Accepting the pain of shutting down big production assets is one of our biggest challenges at KC. But if shutting down helps you fix the problem long term, it is better than continually struggling or never addressing the real root cause at all. But there's more. Because of moving to a fixed and repetitive cycle, bottlenecks and obstacles become more obvious. It helps set priorities. Whatever the problem—machine reliability, material quality, supplier reliability, changeover times—it becomes clear what has to be fixed. So the focus of improvement shifts to what really matters.

In some companies, it has been difficult to accurately assess the direct benefits arising from Lean/R∫S. As improvements have happened, many people lay claim to it being "their work" that has resulted in the improvements. For instance, changeover reductions from SMED exercises, better TPM due to less cancellation of maintenance, improved material flow, and resultant inventory reductions in logistics. That's OK. Lean/R∱S is a foundation—a stable platform that makes the application of all aspects of improvement easier to successfully and sustainably apply. One *should* expect other improvement tools and techniques to be more effective once Lean/R∱S has been implemented and plan changes eliminated.

Impact on Results

This example is from a KC site, Maumelle in Arkansas. It demonstrates the immediate results one gets from a Lean/R∱S 5-day rapid implementation workshop. However, this is just the start, as improvements will continue to grow now that the stable fixed plan is in place. Maumelle has shown output improvements since Lean/R∱S was implemented in every year except one. At the same time, *changeovers per month have increased by a factor of eight*. This shows a level of improvement previously thought to be impossible.

Summary of Improvements from a 5-Day Workshop

Here's a summary of improvements from the 5-day workshop held at Maumelle in September 2006. There were five improvement teams working on different aspects of Lean/R∱S and the barriers to creating a fixed weekly cycle.

Blue team: The team was able to get agreement within the business during the week to a number of "blue" opportunities that involved reduction of non-value-adding complexity: reduce the number of different-sized shipper cases from fourteen down to six; obtain an annual savings of $28,000 due to fewer cartoner changeovers; reduce from two embossing roll patterns to one. Not only did this decrease roll change time, but the team also expects a 2% increase in converting productivity with annual savings of $77,000. *Additional base-sheet changes were made with an annual savings of $359,000.*

Changeover team: With a fixed green-stream schedule grade, changes will occur at the same time in each cycle and in most cases using the same team, resulting in lower ramp-up losses. The repetitive grade changes will decrease change time by half, with savings of $30,000. In addition, having the same crew will reduce ramp-up time, resulting in further annual savings of $44,000.

K2 line productivity team: The team increased throughput on the K2 line, providing annual savings of $178,000. This increased capacity on K2 will enable moving all tub production to K2, freeing the K3 line to produce only refill codes, thus increasing capacity on both K2 and K3.

Rockbusting team on K3 line: Identified and fixed three rocks during the week, resulting in a 15% increase in productivity and providing annual savings of $177,000.

Green-stream team: The team is developing a schedule that includes 65% of current codes in a fixed schedule.

Total annual savings from the 5-day workshop without the impact of economies of repetition = $893,000.

Where did we go from there?

After the original R/S workshop, the Maumelle site was selected as a pilot area for developing increased total value-stream capability. The challenge: *How to use Lean/RfS to connect the improved manufacturing base more directly to customers and suppliers.*

Simply stated, it's all about pursuing smaller batches more frequently produced in a fixed pattern, be it deliveries from suppliers, within manufacturing, or shipments to customers. It is counterintuitive and flies in the face of conventional thinking. Smaller batches help create tighter connections along the whole supply chain and business process. Smaller batches also expose problems that need to be solved. Big batches and high inventories cover up the issues. Problem-solving capability is a critical skill using a standard process to engage team members in solving problems as they arise. It's also where Lean tools need to be applied. Techniques like 5S, visual management, TPM, and SMED are tools that can be applied to address specific problems. These were tools the site had tried to implement before, with some success. However, now the stable repeating pattern, reduction in firefighting, and extra time this gave them made successful implementation of these tools that much easier. It also greatly helped in setting standards for everyone to follow.

So, what has happened since then?

Here is an email to Rick:

From: Painter, Tim
To: Sather, Rick
Subject: monthly report Maumelle

Attached is usual monthly report. By the way, the team believes they are going to set a new volume record again this month!

Regards,
Tim

What we find amusing is the casual way he mentions setting a new record volume—*again*! Improvements include:

- Average changeovers per month:
  2006 = 55
  2009 = 128
  2011 = 400 with some green SKUs four times a week.
- A daily flow (green stream) for finished goods was established from the regional distribution centers back to the local warehouse.
- JIT deliveries from suppliers bypassing the material warehouse.

What results have we seen from this progression?

- Schedule stability increased by 70%
- Parent roll inventory reduced by 80%

- Tub inventory reduced by 75%
- Converting productivity improved by 34% despite the impact of additional grade changes each month
- Process waste decreased by 30%
- Lead time through the baby wipes value stream improved from 120 days to 50 days

It's not just about the hard results numbers. The impact on *people* is important to recognize. During a recent plant visit, Rick asked the base machine operations leader how his world has changed. He replied,

> In our old process, we were constantly chasing the schedule. I had too much of some base sheets and was running out of others, and I personally spent a good amount of my time working with the planners and converting the team on plan changes. Now, the process just works. Converting consumption and the inventory level in the supermarket drives our schedule. Crew members receive triggers to change between grades and simply make the changes following our standard procedure. We now are able to spend much more time on problem solving and improvement. We have reduced our in-process waste to just 1%. In the past we would have never dreamed this was even possible. Now we are looking at how we can cut that in half again. It is a calm and energizing environment to work in.

How can grade changes increase by a factor of eight and productivity improve by 34% simultaneously? This is what Lean/RfS delivers. Stable, predictable processes enable the team to focus on problem solving and improvement, not on firefighting and reacting to changes. The routines allow you to just get better and better. KC has found it never ends—*never*—as shown in the graphs from Maumelle presented in Figures 3.1 and 3.2.

At another company that manufactures baby food, the slide in Figure 3.3 is taken from a presentation to their senior management on the results from Lean/RfS. They identified improvements in five different areas. What we find interesting is that as well as seeing improvements in "hard" numbers such as efficiency, they have also seen improvements in "soft" areas such as employee motivation and supplier relationship. This slide was an overall summary. Specific quantified results are not shown on this slide, as these differed from site to site. These were shown in the full presentation; however, we cannot reproduce the individual site results in this book.

There is another benefit from Lean/RfS that we've seen in several companies that does not show in their KPIs, yet it is a big financial impact. And that is *capital avoidance*. The first example is from Wrigley's, the chewing gum

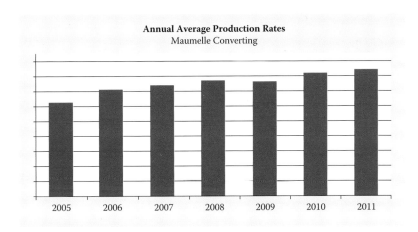

Figure 3.1   Maumelle annual average production rates.

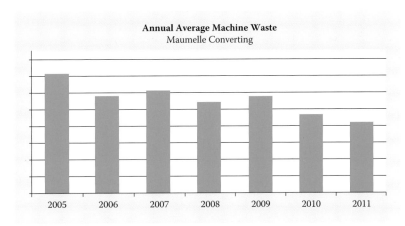

Figure 3.2   Maumelle annual average machine waste.

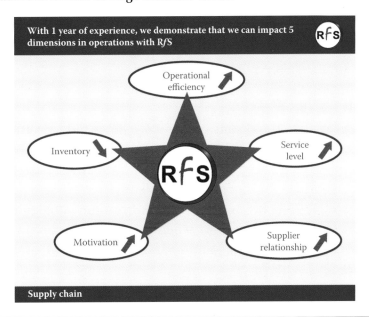

Figure 3.3   Five areas of improvement.

company. Their production director, Alan Richards, told us the about the impact Lean/RfS had had in their warehouse.

> We had a warehouse with 5,000 pallet spaces and it was full. So full, we had to book another 1,000 spaces off-site at a rented warehouse. Now, our finished goods stock for the green items is very low, when it used to be very high. We have moved out of the outside rented warehouse. We have also dismantled racking in our own warehouse because we just didn't need it. It is making space to put in production equipment to make a new product range the company is launching. This is a massive saving for our company, as previously it was working on building a new factory for this product.

Green and Beige People

This is an example where the managing director (MD) took visual management to a whole new level. We can understand why he didn't have red overalls, but we have no idea why he settled on beige instead!

The company was a manufacturer of hydraulic pipes and couplings that produced 2,000 SKUs with a forty-eight-hour lead time to customers, something they were finding increasingly costly and hard to achieve consistently. The Glenday Sieve analysis (see Chapter 2) showed that over 50% of sales came from only nintey-two products, or 5% of the range. The other products were ordered infrequently in small quantities, even as single items.

Armed with this information, the MD decided he needed two types of factory, a high-volume repetitive facility and a flexible jobbing shop. It was not cost effective to build two factories or even to physically separate the equipment inside the existing factory building. The company had to find another way. They worked out what equipment would be needed for the high-volume SKUs in a fixed cycle and the staff required. This equipment was painted green, and all the employees were given green overalls. The remaining equipment was painted beige, with beige overalls for the employees.

Although the equipment and staff were totally intermingled on the shop floor, each business operated totally differently. The "green" factory had fixed hours of work, JIT deliveries of materials, and improvement activities focused on achieving faster cycle times and increased productivity. The "beige" factory had variable hours according to the level of work each week from the variable demand of the small sellers, purchasing materials only when required and running improvement activities focused on flexibility and responsiveness. Product costing was based on the different actual costs incurred in the green and beige factories. This meant that very similar products had quite different product costs, depending on whether it was a green or beige SKU. It did not necessarily translate into different selling prices to their customers; however, the business knew much more accurately what the real margins were—and could make decisions based on that.

They had effectively created two businesses with policies, procedures, and measures appropriate to each. However, the added value, flow, and what was seen as *muda* (i.e., "waste") in each business was completely different. Although on paper the beige factory looked like a more expensive way to make products than the green factory, the reality was that it achieved the flexibility required to meet customer service without causing short-term plan changes and the disruptive firefighting that had impacted the green products in the past. It enabled green products to be made in a totally fixed and repetitive cycle, achieving levels of efficiency, throughput, and lower inventory levels across the supply chain never previously thought possible, at the same time as having the required flexibility and responsiveness for the beige products. Together it resulted in better customer service and overall increased profit margins while still maintaining the range.

An interesting point was how it was decided who worked in the green versus the beige factory. The MD said,

> That was easy. We needed firefighters in the beige team. People who can react quickly and relish the challenge this gives them. The green team were those people who always parked in the same place, people who were known for having their fixed routines and focused on following standards for the way they operated.

A final detail illustrates the mindset issues one can face when moving to Lean/RfS. A visitor to the site congratulated the MD on what he had achieved and asked him how long it would be before all the workers in green were sufficiently trained and experienced to become flexible enough to be given beige overalls. The MD's response was that that would never happen, and the visitor should try and answer for himself why that was. Privately, the MD shook his head, as he knew the visitor had not understood and probably never would!

Policy Deployment

Policy deployment is a key aspect of Lean that helps a company achieve a true Lean transformation. It is not part of Lean/RfS. Once stability has been established, then policy deployment helps to set direction, coordinate activities across the functions, and ensure that business processes are in control. KC used Pascal Dennis* to assist them with implementing policy deployment, to help align and direct the business. We would recommend his book *Getting the Right Things Done* as an excellent explanation of policy deployment and how to apply it in an organization. Policy deployment and its application at KC will be covered in more depth in Chapter 5 of this workbook.

---

* Pascal Dennis is a professional engineer, author, and president of Lean Pathways, an international consultancy. For more information, please visit www.leansystems.org or Pascal's page at Amazon.com.

While employed at Colman's of Norwich, Ian's role was head of policy deployment. The following is a brief case study of how Lean/RfS and policy deployment were combined to create a Lean transformation.

Squash Quosh

At Colman's of Norwich, a UK food and drink company, the principles of Lean/RfS and policy deployment were progressively applied across the whole business. The aim was to increase competitive performance in the market and to improve the financial results. The biggest and most important product portfolio for the business was a type of soft drink. The generic name for this type of product is "squash." It is a concentrated mixture of fruit juice and syrup that needs to be diluted with water before drinking. It is largely only sold in the UK. There were two main brands in the marketplace. Robinsons—Colman's brand—had about 22% market share, while another company's brand, "Quosh," had about 11% market share. A few lesser brands accounted for around 6%, and the rest were supermarket-owned label products.

As Lean/RfS evolved in manufacturing and across the supply chain, Colman's saw improvements in outputs, waste, inventories, and costs. Colman's executive team considered how to leverage this improved performance to take market share from Quosh. To understand the strategy, one has to understand a little about English weather. In the UK, we only get, on average, two weeks of hot sunny weather each summer. We get a few days here and there, but usually there is only one sustained two-week period of good weather. It can happen in June, July, or August, but once it has happened, it is highly unlikely that another will occur that year. During this period, sales of soft drinks skyrocket, and normally manufacturers cannot keep up with demand. Attempts are made to cover this demand through stock building before the summer, but not knowing when these two weeks will occur makes this an expensive strategy.

So drink companies make a compromise. They hold some stock, but not enough to cover all possible sales, as this would be too costly. Traditionally, all drink or drinks' companies always ran out of stock at some point during this two-week hot spell in the summer. It was normal, which is where Colman's opportunity lay. They planned to build capacity so that Colman's sales in these two weeks could be met without stock build *and* so that the company could produce the out-of-sales demand for Quosh, thereby stealing Quosh's market share when they could not deliver. The strategy was code named "Squash Quosh." Some senior executives wanted to keep the strategy confidential. Others felt that everyone in the business needed to understand it so that they could all consider what part they could play in helping to deliver the strategy. The executive team alone was never going to be able to figure out all that would be required to make it happen. A program was developed to inform everyone in the business the strategy goal of Squash Quosh and to ask their help to make it happen. This is part of the process of policy deployment.

Many sessions were held with people from all the different functions in the business. The purpose was to explain the strategy and then ask them to brainstorm what they thought they and the company needed to do to implement it. They were asked to think not what they knew could be done, but what could they do differently—things that had never been done before. The ideas ranged from the ridiculous to the sublime. After much debate, the actions people thought were required were agreed. It would take eighteen months to get them all implemented before they were ready to attack Quosh in the summer during the two-week hot weather spell.

The following are the key activities. Many were considered paradigm shifts from how Colman's had previously operated. But by now, Lean/RfS had shown people that achieving paradigm shifts was indeed possible. In addition, by eliminating firefighting, it had given people *time*—time to think, time to work on doing things differently, and time to make the changes actually happen. They included:

- Harmonizing of bottle shapes so that all flavors had the same shape in each of the three sizes of bottle. This would simplify ordering and packaging inventories and, crucially, would enable the supplier to produce a far smaller range of bottle shapes—a "blue" opportunity realized.
- Having extra molds at the bottle supplier to be able to run more blow-molding machines when required, i.e., in the hot weather peak. It cost, but it gave the molding capacity when it was needed.
- Ensuring that all other material suppliers could meet an increased demand when required. This was easier to achieve than the bottles.
- Moving hourly paid employees onto fixed weekly salaries. Then, adjust their hours of work to match the seasonal peaks and troughs. This meant that they worked longer in the summer and less in the winter but were paid the same amount each week. This is called *annualized hours*. It was a major change that took a lot of negotiation to achieve. Two things helped to make it a reality.

  Firstly, because we had literally "every product every week" in fixed eight-week cycles we knew exactly how much needed to be made each week. Therefore, when the right amounts had been made, the factory had been cleaned, and all 5S standards had been met each week, the people could go home. Annualized hours are something that many companies have tried—and failed—to implement. In our opinion, implementing annualized hours in an organization where batch logic is used to calculate what needs to be made is bound to fail. Having a different plan each week and then changing it is akin to moving the target, thereby giving the operators a constantly moving target that means they are nearly always at the bad end of the deal. Doing annualized hours in a Lean/RfS environment meant that *both* the operators and the company were beneficiaries of any increases in performance.

Secondly, it also meant that the need to employ temporary workers in the peak season was eliminated. This was seen by the permanent operators as positive, as line performance always suffered when temps were used.

- Analyzing the retail supermarket stores using the Glenday Sieve to identify which were the "green" stores that accounted for 50% of sales in the UK. Then, having the salespeople build strong relationships with the "green" store managers, so that when the hot period came, they were in a good position to persuade them to fill the empty Quosh shelves with Robinsons Squash.

- Engineering each of the five bottling lines and processing equipment so that they could all produce the 750-ml bottle in whatever flavors needed.

- Completely changing tactics during the hot period. There were twenty-eight SKUs that were usually made every week in a fixed Lean/R/S eight-week cycle—Every Product Every Week (EPEW). The plan was that when the weather forecasters (Colman's had their own specialist people for this) predicted that the two-week hot period was imminent, production would switch from EPEW to only making the two biggest-selling SKUs: 750-ml orange and 750-ml apple and blackberry. We would go from economies of repetition to economies of scale and keep all lines running at maximum output on only one variant, so there were no changeovers at all. We had gambled on the likelihood that when the stores started running out of Squash in the hot period, store managers would be happy to take any flavor just to have product for their customers. A bonus would be that the only SKUs available would just happen to be the customers' favorite choices.

- Communication, communication, communication—keeping people informed and aware of all that was going on and having visual tracking of performance in every function and clear visibility of the actions required. This is all part of successful application of policy deployment. Some were worried that all this communication would reach people in the competitor company, Quosh, so they wanted to suppress it. Two arguments won them over. Firstly, Robinsons was number 1 in the market and Quosh number 2. Don't you think they already know we are targeting their market share? And secondly—much more persuasive—even if they knew what we were doing, what could they do about it? Also, many felt that people in the competitor company were likely to dismiss it as too far-fetched to believe. In any event, people didn't need to worry. Everyone at Colman's was so excited and committed to the strategy that no one was going to tell.

- Installing a temporary bottling line to satisfy peak demand. This final key action was the most difficult to achieve, as it stretched some people's conventional thinking to the breaking point, in particular the people at the head office. Despite all the efforts in manufacturing, we could not achieve the outputs needed to produce the amounts we had calculated would be required to equal both companies' expected sales across the hot period. We had to install another bottling line just for this period. Head office found the idea of a multimillion investment for just a few weeks of production somewhat crazy. Luckily Colman's MD was a charismatic and persuasive individual who managed to get agreement for it. The question then was how to staff this line during these few weeks. When we raised it in the weekly policy deployment communications, the answer came back almost immediately. People from other functions in the business volunteered to be line operators during the hot spell. A training program was developed to give them the right skills and safety awareness.

As the summer approached, the excitement in the company grew. Everyone was asking, "Is the hot period coming yet?" It was one big team working together on a common goal: to take market share away from Quosh and to grow the Robinsons business. But each week the forecasters said "no." But then they changed to "it's coming." The atmosphere became electrifying—it was truly an awesome experience. Production was switched to only two products, suppliers were asked to increase their outputs, and salespeople went to each of the identified "green" stores to wait for Quosh to run out of stock. As soon as this happened, they spoke to the store manager to say they could fill the shelf with Robinsons Squash that day. Of course, the store managers agreed. The sales-people then moved onto the next-biggest stores, and so on. During the two weeks, Quosh's market share dropped from around 11% to less than 1%.

Once we had the shelf space, we were not going to give it up. The strong relationships the sales team had built up—and the fact we had supplied when Quosh could not—meant that they were willing to increase shelf space for Robinsons at the expense of Quosh. Plus, another element of the strategy that had been kept secret came into play. The performance improvements coming from Lean/R/S had increased profit margins. The salespeople were now able to give the retailers better terms using some, but not all, of this extra margin to encourage them to give the shelf space to Robinsons.

The result was not what had been planned. The plan was to take possibly 4% or 5% of Quosh's market share. With less than 1% market share, Quosh soon folded, as it could not recover its fixed costs. The brand was withdrawn from the market. Back at Colman's the reaction was amazing. What had become a much better place to work with greater teamwork and recognition now took another step forward. It was *the* place to be. Sales doubled per employee, market share grew to just over 40% (we also took some supermarket-owned-label share), and profit margins went from 12.8% to 17.3% in a sector where margins were typi-cally 5% or 6%.

Unfortunately, there is a sad footnote. The main product ranges of Colman's parent company were in household cleaners and disinfectants on a global scale. A largely UK-based food and drinks division was "off strategy"—even though Colman's had the best margins by far in the total organization. The parent company appointed a new CEO who did not understand Lean/R∫S. He also saw an opportunity to buy a major disinfectant brand in the United States, where the parent company was weak. The cost was too high to fund without raising capital by selling something. That something was Colman's. It was considered "off strategy" as far as the CEO and the parent company were concerned. No company wanted all the Colman's brands, so they were split into two, with one company buying the food brands and another the Robinsons brands. These companies wanted the brands—not the companies or the assets. They did not look at how the company operated or why Colman's had achieved such great success.

The new owners managed to destroy what had been built in a short time. They did this because they did not understand Lean/R∫S and already had a plan to "save" the costs of buying the brands through "efficiencies" achieved by transferring as many activities as possible to their existing organization. The result was many redundancies at Colman's, and most activities eventually were moved to the sites of the new owners.

Changes at the top of the organization are a major danger to the continuation of Lean/R∫S in a business, especially if the people concerned do not understand it and still think the conventional wisdom of batch logic is right.

Stages in a Lean Transformation

At times, even the drivers of change find it hard to believe the scale of what can be achieved with Lean/R∫S. It happened to Rick. After Lean/R∫S had been implemented at several sites in KC—moving to weekly production for the green items—Rick said to me, rather pointedly,

> Getting to weekly cycles has been a major challenge, but we have demonstrated it. We now know we can do it despite the obstacles. But Ian, don't you ever, ever, *ever* talk to me about getting to the next step of leveled production "every product every day." It is *impossible* in our industry.

A while ago Rick sent out an email to his colleagues at KC. Here it is.

Subject: Money Falling From the Sky

I usually sleep very well and rarely ever get up in the middle of the night. However, this week, it happened on Tuesday night when I woke up at 2:30 a.m. to the point where I needed to jot down a few notes to get them out of my mind so I could get back to sleep.

I was thinking about Every Product Every Day = EPED. I've thought a great deal about this during the past three years after I was first exposed to R∫S with Ian Glenday. Getting our products to single-piece flow is very hard for me to imagine at all, but I have believed for a while now that we could eventually get to EPED. I thought this would perhaps take five to ten years to accomplish. Now, as we gain a deeper understanding and priority for lead-time reduction, I believe we will and need to get to EPED much sooner. How much sooner is the reason I'm sending out this e-mail to you (my EPED manifesto).

Imagine if we could produce Paper Towels EPED at the Jenks facility. Below is a summary list for some of the benefits we would see, just at this one site. I believe there are many more than these, but I wanted to give you a taste for the possibility.

1. The strategic planner for Towels ran a model of EPED inventory for finished goods. We would be able to reduce inventory by seventeen days.
2. The resulting benefit to reduced storage and handling costs would be nearly several million dollars.
3. WIP parent roll inventory could also be significantly reduced, and this space could be used to reduce external storage in our network, probably also worth several million dollars.
4. Since we would be making EPED, a high percentage of Towel production could go directly into trailers for customer shipments, further reducing handling expense.
5. With an EPED cycle, we could arrange supply deliveries on a daily basis (maybe even more than once a day) and substantially reduce Finishing Supply inventories.
6. Direct flow paths with milk runs could be set up for material flow, and we would have substantially reduced transport waste.
7. Parent roll waste could be reduced, as many rolls could go right from the OLU [off-line unit] to the winders.

I believe a *very* conservative estimate for the financial benefit for all the above to exceed $20 million *not* including the balance sheet benefits of lower inventory and other benefits from cutting lead time for these product lines in half.

We'll need to solve many problems and build a lot of capability to get to this point. I'm very much looking forward to this next step of leveling, and I can already envision flow workshops and Kaizen events for this business.

Hopefully, this gives you a taste of what EPED could do for us. I now throw out the EPED challenge.

Rick

What Rick once saw as impossible, he now sees as both desirable and doable—albeit with some big challenges. Rick has changed his view of what is possible.

We have found that many people find it difficult to imagine what the stages are in a Lean transformation—how, from starting with Lean/RfS in

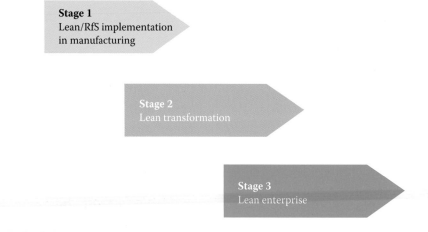

Figure 3.4   Lean transformation stages.

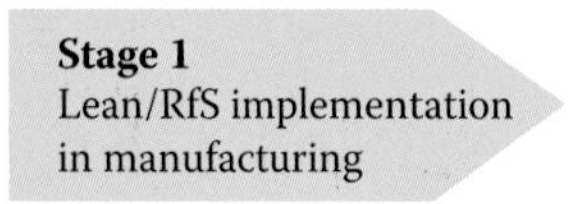

Stage 1
Lean/RfS implementation
in manufacturing

- Manufacturing implementation of Lean/RFS

- Lean learning – build expertise of the "early adopters" and start education for leadership

- Increase rapid improvement capability through workshops and common problem-solving methodology

- Demonstrate benefits to support expanding Lean-RfS
  Step change rate of improvement for throughput, conformance to plan, quality and inventory reductions

- Establish systems, structures, and KPIs to institutionalize Lean-RfS in manufacturing management

- Initiate Stage 2 in parallel

Figure 3.5   Lean transformation Stage 1.

manufacturing, it might develop into a company-wide change, delivering step changes in market share, customer service, sales per employee, and profit margins while increasing employee satisfaction and motivation. We developed a model to demonstrate what the stages could be. It's not a project plan, and it does not have a timeline. We considered it as a route map to help show what the steps in the journey could be.

The model has three stages, as seen in Figures 3.4–3.7. Each stage has a number of characteristics or elements to it. For Stage 1, these are as shown in Figure 3.5. Stage 2 (see Figure 3.6) is about progressive expansion of Lean/RfS concepts across the whole supply chain and throughout the business. In addition, policy deployment needs to be implemented. Stage 3 (see Figure 3.7) is where Lean becomes embedded in the whole organization because it has become a demonstrated competitive advantage.

Stage 2
Lean transformation

- Lean leadership learning through doing

- Application of strategy/policy deployment

- Build problem-solving culture in the business

- Extend Lean/RfS across the supply chain
  key customers and suppliers

- Manufacturing progression to the next steps in leveling
  100% of products in the cycle (true EPEC) and faster cycles

- Leverage Lean/RfS capability for business benefit
  Market expansion and/or major cost savings

- Extend systems, structures, and KPIs to institutionalize Lean/RfS
  beyond manufacturing

Figure 3.6   Lean transformation Stage 2.

Stage 3
Lean enterprise

- Broad and deep cross-business engagement to drive out process variability
  and institutional waste

- Fast, flexible ability to supply to demand

- Demonstrate ongoing dramatic improvements in lead time reduction,
  process capability and business results

- Systems, structures, and KPIs in place to manage across value streams

Figure 3.7   Lean transformation Stage 3.

Kimberly-Clark still has a way to go to get to this stage. It may never get there. It certainly won't get there if it does not at least create the vision that it *can*.

Summary of Chapter 3

We have seen some amazing results from Lean/RfS: not only improvements in performance, but just as importantly—if not more so—in changing people's behaviors, attitude, and motivation far beyond what was previously thought possible. This has been pleasing, but has also challenged us to think, "How far can this go? Just how much better can it become?" Chapter 3 has attempted to share some of these results to give the reader an appreciation of what is possible. We have also tried to show how achieving a true Lean transformation takes much more than just Lean/RfS. It requires policy deployment as part of Lean leadership, and that's something that is hard to successfully accomplish. This, and the lessons learned, will be covered in more detail in Chapter 5. However, the main focus of this workbook is Lean/RfS—*getting the right pieces in place to stop the firefighting and create a stable foundation on which one can then build a truly Lean enterprise.*

Chapter 4

The Lean/R*f*S Corner Pieces

Changing from Batch to Flow

When solving a jigsaw puzzle, people usually find the corner pieces first and then put them into place. The job of the corner pieces of Lean/R*f*S is to stop the firefighting caused by batch logic through:

- Implementing a fixed plan for green items
- Using a buffer tank to absorb normal sales and production variability
- Integrating Lean/R*f*S into existing planning and production management processes

These result in a stability that provides the foundation for sustainable improvements and increased performance across the entire supply chain and in all business functions, making it possible to achieve better product quality, higher market share, increased margins, and happier employees. It also gives people *time*—time to think, time to investigate root causes, and time to implement standards. Sustainable improvement is very difficult to obtain when there is a different plan each week, when the plan keeps changing, and when firefighting and symptom solving are the normal way of operating. This is the situation we find in most manufacturing companies because they are using planning systems based on batch logic or economic order quantity (EOQ), either of which will guarantee a different plan every time, combined with unnecessary levels of variability.

Batch Logic Is Bad

An analysis that reveals much about batch logic is to consider what was planned and produced for the highest-volume products over a six-week period. Table 4.1 shows the results of such an analysis produced for a well-known, highly respected, and very successful company.

Table 4.1

		Product A	Product B	Product C	Product D
Week 1	Plan	40,000	72,000	72,000	81,000
	Actual	94,908	121,000	103,000	57,000
Week 2	Plan	36,000	84,000	63,000	69,000
	Actual	0	73,000	0	135,000
Week 3	Plan	96,000	72,000	132,000	42,000
	Actual	108,000	80,000	108,000	47,000
Week 4	Plan	84,000	66,000	69,000	69,000
	Actual	82,000	0	83,000	102,000
Week 5	Plan	72,000	84,000	51,000	69,000
	Actual	113,000	91,000	58,000	69,000
Week 6	Plan	108,000	90,000	45,000	60,000
	Actual	58,000	130,000	57,000	90,000
Total planned		436,000	468,000	432,000	390,000
Total produced		455,908	495,000	409,000	500,000
% Plan/Actual		105%	106%	95%	128%

There are several interesting points to consider here. Firstly, this company already makes their biggest volume—green items—every week. So there are no technical barriers, such as long changeovers, to be solved before the implementation of a fixed weekly green-stream cycle. Secondly, the quantities planned each week vary considerably, and a different plan is issued each week. This example concerns an industry where the consumer base is very brand loyal. As a result, the actual consumer demand is exceptionally consistent and predictable. Thirdly, the difference each week between what was planned and what was made is truly amazing. The senior management in the company was genuinely shocked when they saw the figures. However, over the whole six-week period, the total planned and made was relatively the same. In fact, overall a little more was made than planned, and the factory management saw this as good. It meant that their efficiencies and fixed costs looked favorable.

This situation is not unusual. Here's an example from a Scotch whisky producer. The main customer of the Scottish bottling plant is their U.S. affiliate company. The affiliate's orders over a six-week period are shown in Figure 4.1.

They ordered every Thursday—good routine and repetition. However, the orders varied between 4,000 and 24,000 cases across a period where there were no promotions and demand was stable. The planner received these orders, and Figure 4.2 shows what he planned to produce each week.

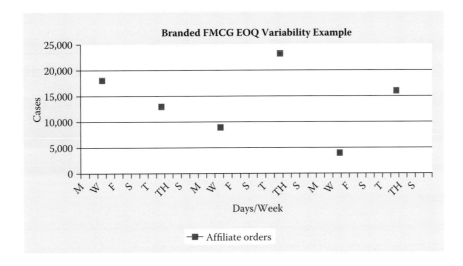

Figure 4.1 Variability of supply chain.

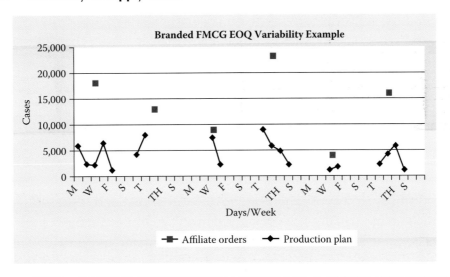

Figure 4.2 Variability of supply chain with production plan.

Every week is different, with different quantities on different days of the week, because the order volumes varied. Figure 4.3 shows what the factory actually made.

On a week-by-week basis, each week's production was more or less what was planned. But actual daily outputs bear only a fleeting similarity to what was scheduled! Overall, this supply chain was delivering what was required, and the company was proud of its high customer service levels. But the degree of variability—in affiliate orders, production schedules, and actual production—was amazing. These are just two examples of what happens when using batch logic.

Think about the following points.

What is the risk of errors being made? Are there errors that could potentially impact product quality or employee safety? What is the effect on suppliers? How much are they being messed around? What is the likely effect on their ability to supply reliably? How many tasks and activities are required to make such a

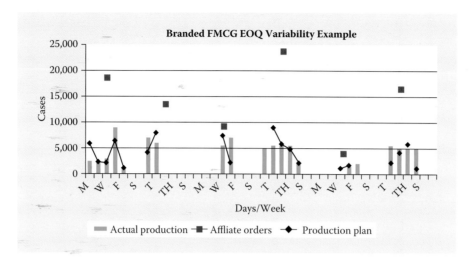

Figure 4.3   Variability of supply chain and actual production.

situation work? How is this driving fixed-cost overheads higher than required? The risks and costs of operating in this fashion are far higher than they need to be. Batch logic results in a different plan every time. *Batch logic is bad.*

What Is "Responsiveness"?

People sometimes claim that having a different plan each time and then being able to change it after it has been issued means that they are "responsive," the implication being that this is responsive to customer demand changes and therefore a good thing. However, when one looks at real demand, especially for green items, their demand usually has only a relatively small percentage of variability. Responsiveness, or doing something different each week and then changing it when there is basically no significant change in demand—as demonstrated in the previous examples—increases costs and the risk of errors. This is not responsiveness. *It is madness brought about by the way batch logic works.*

Real responsiveness is being able to detect when actual demand goes outside of the predicted and normal levels of its sales variability. *Only then* should the plan be changed to meet the new situation. When there are no significant changes to real demand, one should be able to keep the production plan—and therefore the whole supply chain—stable. This should not be a surprise to anyone. After all, it uses exactly the same logic as statistical process control (SPC), and we have applied that in production to control things like weight or fill levels for years.

An illustration of the impact on plan changes when switching from batch to flow logic is shown in Figure 4.4. It shows the number of changes made each week after the plan had been issued in a factory producing biscuits.

Weeks 1 to 10 illustrate when the company was using batch logic to plan with. On average they changed the plan three times per week, and some weeks up to eight times! Week 11 onwards illustrates what happened after management had moved to flow logic. It was Week 38 before they made a

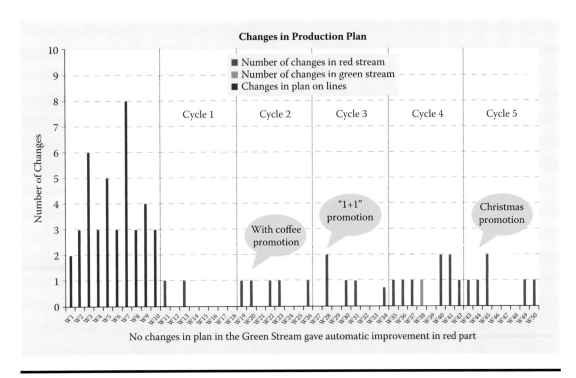

Figure 4.4 Changes in production plan.

single change to the green stream after the eight-week cycle plan had been issued. They went six months without a change when previously they averaged three times per week! During that time they had three promotions, so demand was not constant, yet the plan remained stable. Previously, the company had been very "responsive," but not to changes in actual demand. This responsiveness—caused by batch logic—only created increased noise, risk, and costs. The level of stability under flow logic really surprised them. They never expected it to be quite *so* stable. However, what shocked them was the reduced level of changes needed in the red stream—those smaller volume products whose demand *is* more variable. The company saw a reduction in these as well, because now they were only reacting and being "responsive" when actual demand had genuinely changed. And this happened far less frequently than people previously realized or recognized.

What many people claim as "responsiveness" in their supply chain and in factories is just wasting resources due to the fundamental flaws of batch logic. Responsiveness? I don't think so!

Let's look at a large global branded FMCG company, which shall remain nameless for the sake of its reputation, where the management considered themselves very "responsive." Rest assured their case is far from unique. We asked the Global Supply Chain vice president to map what happened to the weekly production schedule at one of their factories after it was issued. How many copies were made? Where did they go? What other plans were generated from the schedule? How many times a week did the plan change? Figure 4.5 shows the map he drew.

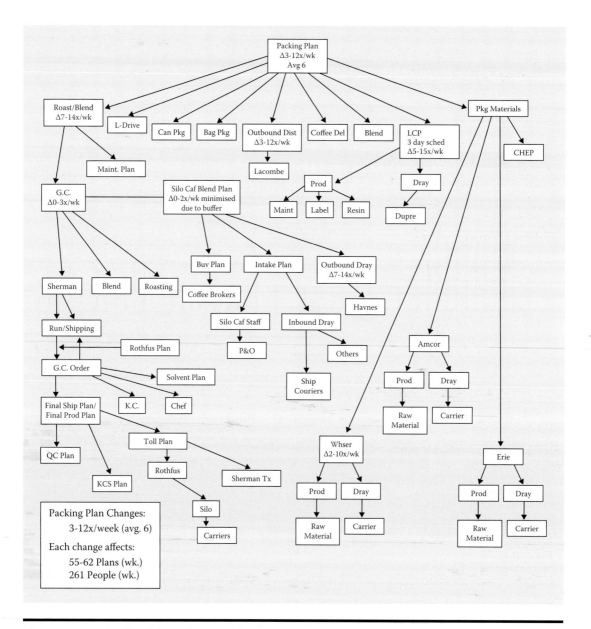

Figure 4.5 Distribution of the weekly production schedule.

It shows that fifty-five to sixty-two copies, or derived plans, of the weekly production schedule were delivered to different areas. Then, it changed on average six times a week! Was there *any* chance that everyone would be working to the same plan? This is a highly successful and well-regarded organization equipped with the latest ERP computer systems.

But it gets worse. This company prided itself on building flexibility into its manufacturing facilities. As the product progressed through the various stages of manufacturing and packing, people were able to change routings through the factory to switch from one piece of equipment to another. There was a good reason for this—or so they thought. It was "to meet the demands of our customers when the demands change"—what many would call "responsiveness."

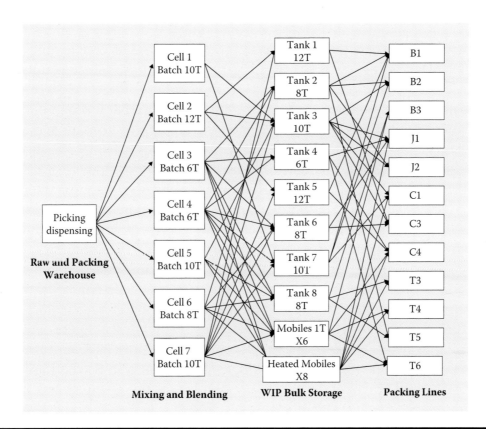

Figure 4.6 Plant layout plan.

But when we asked the company to keep a simple check sheet of why routings were changed, the biggest reason given was issues within their own operation. The real reason the company placed so much emphasis on manufacturing flexibility was to cope with problems arising in its own production process: "If there's a problem, like a breakdown, switch from one piece of equipment to another." Figure 4.6 shows the pipelines between each stage of manufacturing for a factory making liquid products.

Now couple this "flexibility" to the schedule map. Is it any wonder this company found it hard to create and keep to stable plans? The result was a supply chain characterized by endless firefighting, chaos, and stress for everyone involved, and a business operating with higher costs and lower efficiency and effectiveness than it needed to.

Calculating the Schedule

Generally companies use some sort of computer program to plan their production. It's often a program bought from an IT supplier or on a spreadsheet developed by someone within the company. Either way, the aim is to get the computer to create a plan. This plan is based on certain criteria built into the program. It is impossible to program all the potential variables that may have an impact on the plan; for example, a key worker who has specialist skill in running a certain product being on holiday, or changeover times varying depending on what time of day or night they happen to fall, caused by the number of people and their skills available to do the changeover. A computer cannot produce a plan that optimizes all the possible variables because, by definition, these variables change. Computers can only work according to the way they have been programmed.

Humans, on the other hand, are very good at taking into account and adapting to changing situations and multiple variables. For this reason, calculating the fixed green stream should be carried out by humans, not computers. It needs to be completed by a team of people representing production, planning, management, and sales. All plans are trying to achieve the optimum balance of conflicts; there is no "perfect answer." So each area needs to present its requirements and optimum solutions for the plan. Then, through debate and analysis, people mutually agree on the optimum overall plan. This is a time-consuming process at first, taking longer than a planner working alone to determine a weekly schedule would take. However, remember the plan for the green-stream items will be the same each week for (usually) the next eight weeks. So the activity of determining this weekly fixed schedule for green items only must be carried out once every eight-week cycle. With flow logic, it is *people* who do the thinking and decision making required to agree on the plan for the green stream, not a preprogrammed computer planning system.

After the very first green-stream schedule had been agreed upon at KC, the scheduler, Walter Nawrot, made a telling comment. "It's no longer my schedule; it's our schedule," he said, which neatly summed up the change in approach, understanding, and buy-in that had happened. With each cycle, the time taken to plan the green-stream schedule gets much quicker. For one thing, people tend to try and make as few amendments as possible to the green-stream sequence. They start with the current schedule and only change things they feel are essential—like demand volumes and output rates.

This workbook is not intended to be a detailed "how to" manual on determining a fixed green-stream schedule. Getting the right people together—ideally in the same room but, if not, via a conference or video call—is the key requirement. They will have the skills and experience to analyze and agree upon a green-stream schedule. This approach results in greater ownership and commitment from everyone to make it work. If a more detailed and specific explanation of how to calculate this schedule is required, please refer to the workbook *Breaking Through to Flow* by Ian Glenday.

Out of all the points to consider when calculating a green-stream fixed cycle, here are a few that are fairly common when moving to flow logic.

A big consideration is whether to run to quantity or time. Batch logic always plans to quantity, to produce a certain amount. But, as the previous examples have shown, production people do not always make the quantity planned on the day it was planned to run. This causes variability and the inevitable disruption. With flow logic, the alternative is to *run to time*. There are two main reasons for this.

Firstly, to offer the best possible environment for economies of repetition to
  occur. People click into routines much quicker if they are time-based—it's
  the same time every cycle. They know when things are supposed to happen.
Secondly, to have certain things happen at their optimum times, for example,
  changeovers, so that they happen when the right people are available.
  Running to time also emphasizes the fixed nature of the plan. The idea of
  running to time is a paradigm shift in thinking at first for many people,
  but experience shows that in most factories it is possible to achieve. It may
  require a fixed window of time, say ± thirty minutes, rather than fixed to
  the minute. Running to time versus quantity is covered in more detail in
  Chapter 6, including how it will help to increase motivation.

How to sequence within the cycle to produce certain products is another
key consideration. It is best to let the operators decide when and in what
sequence they prefer to run the products, as they will have a better understanding of this than anyone else. There are some common preferences seen in
different factories. For example, in 24/7 operations, people will always want to
run green items over the weekend in order to run the products that are generally easiest when there are usually fewer resources around to help with running
problems. In five-day operations, people typically want to start and finish the
week on greens, to get the plant off to a good start on a product where the
machine settings are better understood, and then finish the week on standard
machine settings.

Having decided where operators prefer to run the greens, the schedulers can
now use the remaining time to plan each week around which, and how much
of, the "reds" to run. This part of the schedule will be different each week, as the
sales variability of reds is greater. Also, it would be difficult to make all the reds
every week due to changeover times and production losses on small volumes of
product. Of course, one should question whether the reds should be made at all.
This will be covered in more detail in Chapter 5 with a very practical approach
that can be implemented once the green and red streams are operational and
that will help people to understand the real cost of reds to the business.

A misunderstanding that can occur is the difference between "gap" and
"white space" in Lean/RfS terminology. Gap is built into the schedule to be used
if things go wrong and insufficient green product is made. This might be overtime or running the product on a backup line or even producing at a copacker.
It is contingency planning to apply if ever things go wrong with the greens.
If one cannot run to time and one has to run to quantity, then "gap" time is
sometimes planned before tasks that one wants to happen at specific times,
such as major changeovers. The reason is to allow some time to compensate for
variable outputs and still be able to hit the fixed time of major activities.

White space is specifically a result of economies of repetition occurring. It is
an indicator that things are improving. If greens are run to time, then white
space will generate more output in the same time. As a result, OEE (overall

equipment effectiveness) will also go up. When running greens to quantity, the line stops when the planned amount is made. So white space is time. This is because with Lean/RfS, the next production run is *not* pulled forward. To do so would disrupt the drumbeat in the rest of the supply chain, plunging it back into a firefighting mode. In both cases, the white space generated should be reabsorbed back into production output rates at the re-sieve meeting when the next cycle schedule is agreed upon.

If a factory operates 24/7, then creating "gap" becomes more difficult than just agreeing to running overtime at the weekend—the most common "gap" allowance in five- or six-day operations. This is why a backup line or copacking may be an option. It is not always possible to have "gap" to help protect the green stream, but that should not stop implementation; it just means that people need to be very clear about the rules in managing the buffer tanks, which is covered in the next section.

Buffer Tank Calculation and Rules

With batch logic, the planning system is trying to answer two questions for each product:

◼ When to make it
◼ How much to make

This is based on the reorder point, plus safety stock level and minimum order quantity. There may also be a maximum stock level to consider, but not always. The most significant figure is the reorder point, as no production will be planned by the system until this is reached. It is the "trigger" that determines something needs to happen to initiate production of a product. Therefore, making sure it is correctly calculated is crucial. Get it wrong and you will either make it too soon, thus increasing inventory, which is a *bad thing*. Or, you make it too late and potentially give rise to a customer service issue, which is a *very bad thing*.

With flow logic, the calculation of how much to make and when to make it each week has already been done by people working together on calculating the green-stream schedule, as described in the previous section. Therefore, with flow logic, the planning system does not need to "plan production." What is required is a reporting process to highlight when stock goes outside the buffer tank limits—as a warning to look at it. Being outside the limits is not necessarily bad, and certainly should not be an automatic "trigger" to change the plan. The limits are merely the point that indicates it should be highlighted to the planner and that they should investigate why this has happened. For this reason, the buffer tank limits are not critical. They do not "trigger" action. Products will continue to be made each week at the same time and in the same quantities, so there is a "built-in" stability in the flow logic.

The setting of the limits has more to do with judgment than mathematical calculation. Some people are naturally cautious, and so tend to set the limits wide. Others are more aggressive in wanting inventories to be reduced, so they set lower limits with less safety stock. Many planners find it difficult to accept that these limits are not critical after preparing years of detailed analysis to determine the optimum setting for the reorder point.

A Lesson in Setting Accurate Buffer Tank Limits

When the first R/S implementation workshop was run at KC Family Care, the planning director, Jayne Kelly, took on the responsibility of calculating the buffer tank limits. She used various statistical analyses to set the best and most precise limits according to customer service objectives, sales forecast accuracy, production output variability, and inventory targets. The limits were calculated to several decimal places of accuracy and took her many hours of work.

Then, eight weeks later, the second R/S implementation workshop was run at another site. The planner given the task of calculating the buffer tank limits for this site rang Jayne to ask her which statistical analyses were the best to use to calculate the buffer tank limits. Jayne's answer was "just set them at two to four weeks—that'll do." She had realized from the first implementation that the production happened reliably, and sales did not actually vary that much each week. As a consequence, the inventory tracked at a much more steady and predictable level than she had anticipated, and well within the limits she had calculated. Jayne had learned it really wasn't that critical for the green items.

To determine the limits, one needs to look at actual sales variability: how much the weekly sales, in percent terms, vary from the average weekly sales. If the sales are seasonal or have periods of promotion, one needs to break the year up into representative sections or blocks and then use the sales variability in these blocks to set the limits. Typically, green-item sales do not, at a percent level, vary that much. There is often a perception they have high variability, but when the analysis is done and a graph of percentage weekly sales variability is drawn, it usually shows a lot less than people believe.

Here is an example from a very different environment from manufacturing that will help illustrate this lack of variability in green-stream items versus the perception that people have of it.

Having implemented R/S in a hospital operating theater, it was suggested that we look at their accident and emergency department (A&E). One of the fundamentals of R/S is having a relatively predictable demand for the green items, which in the context of an A&E would be the number of patients with the same or similar injuries. The response of the doctors was that demand in A&E was highly variable and unpredictable, precisely because accidents and emergencies were not predictable. They lived it every day and they just *knew* it was highly variable. However, when the data was looked at, certain patterns started to emerge. It was true that demand on different days of the week and at different times in the day was very variable. However, demand on any particular day and time was amazingly consistent for the "green" injuries of patients. One needed to compare Wednesday afternoons to Wednesday afternoons or Saturday nights to Saturday nights. We realized that the doctors and other staff working in A&E had not seen these patterns. Why? Because they worked shifts: sometimes days, sometimes nights, and with different days worked by each person each week.

Consequently, they were presented with different demands each time they began work. To be able to connect the demand on a particular shift to that same shift week after week and so be able to see these patterns was very difficult, especially when a particular shift they had last worked might have been weeks ago. It was just impossible for them to see the patterns, so the perception was that demand was highly variable and unpredictable. It is often the same perception in manufacturing companies. Demand is seen as highly variable. Yet analysis of green-item demand—at a percent level—usually shows demand fluctuates by less than ±50% weekly, meaning that plus or minus half a week of stock cover is sufficient to cover this level of demand variability. Most people take a cautious approach at first and thus set limits at plus or minus a week of sales cover. *The buffer limits are not that critical!*

After determining the width of the buffer tank limits, the next decision is what to set as the lower limit. Between the lower limit and zero stock is what is known as safety stock. With batch logic, safety stock is calculated based on various considerations like production lead time and reliability, sales forecasting accuracy (or rather inaccuracy!), and service-level targets. The safety stock is there because there is a set amount of time, called lead time, between production triggered by the reorder point and when the product is actually received.

However, the actual time it takes varies. One gets differences in production outputs versus plan and actual sales versus forecast; hence, safety stock copes with these differences. With flow logic, the production is fixed, so there are no production lead-time issues. The buffer tank limits have been set using the known sales variability, so that is taken care of. This means that the safety stock is only required to deal with truly totally unpredictable events and issues. How does one calculate a number for totally unpredicted events? Well, it's very difficult. The safety stock with flow logic tends to be more a judgment than a statistically calculated figure. People find this difficult to accept at first, having been used to calculating safety stock based on certain conventional batch-planning logic algorithms. Hence Jayne Kelly's response "two to four weeks will do." When determining for the first time what the safety stock should be, most people decide to set it at whatever the current level is. It is a low-risk option and makes a lot of sense. It can be reset—usually at a much lower level—when one has gained some experience of running Lean/RfS and they have seen the level of stability it creates.

After there is agreement on the buffer tank limits, it is important that two reporting processes be put in place. Firstly, a tracking report to identify any items outside the limits must be established. This is the report planners should use to identify when they should investigate why any item is outside the limits. If they

Table 4.2

	11/27	12/4	12/11	12/18	12/25	1/1	1/8	1/15
STING8RR	OK	OK	OK	OK	OK	OK	OK	OK
STING1CAS	HIGH	HIGH	OK	OK	OK	OK	OK	HIGH
STING3RR	OK	OK	OK	OK	OK	OK	OK	OK
STING12RR	OK	OK	OK	OK	OK	OK	OK	OK
STING15RR	HIGH	HIGH	HIGH	OK	OK	OK	OK	OK
STING1MR	OK	OK	OK	OK	HIGH	HIGH	OK	OK
STING1RR	OK	LOW	OK	OK	OK	OK	OK	OK
STING24RR	OK	OK	OK	OK	OK	OK	OK	OK
STING2CAS	OK	OK	OK	OK	OK	OK	OK	OK
STING2MR	HIGH	HIGH	HIGH	HIGH	OK	OK	OK	OK
STING6RR	OK	OK	OK	OK	OK	OK	OK	OK

are inside the limits, i.e., OK, then planners should avoid the temptation to "just take a look at the inventory levels." Table 4.2 is a typical example from KC.

When the stock is outside the limits, some companies decide to show by how much in days of cover. This helps the planners decide whether or not any action is required. Usually, if it is only a few days, the planner will wait a week to see if it swings back into the limits. This is because experience has shown that it usually does!

The second report is required to help determine the buffer tank limits in the next cycle. As such, it only needs to be produced for the re-sieve meeting once every cycle. It is the equivalent of a run chart in SPC, i.e., it shows the actual inventory in relation to the limits over the whole cycle (see Figure 4.7).

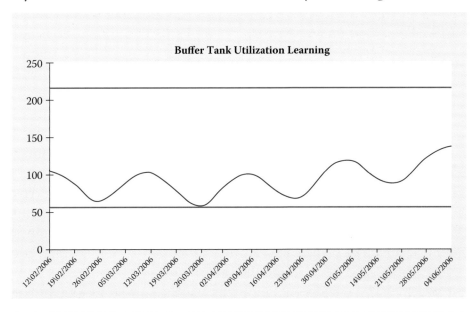

Figure 4.7   Actual inventory run chart.

It is quite easy to interpret the graph. In this example, the upper limit was set too high, but it can be brought down. Although the lower level has been exceeded once by a very small margin, there is still quite a lot of safety stock—fifty units when the average weekly sales are around seventy-five units. The lower limit could be reduced by, say, twenty-five units, representing a saving of 2.3 days of inventory, which in today's economic environment may be a welcome cash and space benefit to the company.

More important than determining the buffer tank limits is deciding the rules. These are *not* guidelines, but rules everyone will follow. Lean is as much about following standards and disciplines as it is about flow. For flow to work, one must set down rules to ensure that people understand what they should be doing, to help them make the transformation from operating in a batch-logic fire-fighting situation to a calmer flow-logic one. The rules should be agreed as part of setting the green-stream schedule and involve the same participants representing the manufacturing, supply chain, and sales functions of the business.

The debate on what the rules should be is an important part of the process so that people understand why they have agreed to the rules. For this reason, we do not recommend a set of rules as being the "right ones." Some companies have only a few rules in order to keep it simple as well as not being too prescriptive, believing they have a good understanding of flow and can work well together to make it work. Other companies have many rules in an effort to cover every eventuality in addition to defining strict criteria to prevent people "bending the rules" and going back to old ways of working. Here is an example of simple rules. This company has a very clear view about the importance of protecting the green stream.

Statement of Fact Rules
- We will not change the plan on green stream to compensate out of stocks.
- Red products cannot run in the green stream, but green products can run in the red stream.
- If we are running behind the plan in coating, we will stop at the planned time and move to the next product in the cycle.

The next example is one where there are more rules. Each company needs to decide how many rules it needs to ensure that the disciplines of the green stream are understood and followed.

Green stream rules
1. Establish frozen cycle of eight weeks.
2. Green: production is measured with time.
3. Red: production is measured in quantities.
4. Don't change the plan in case of breakdown.
5. Don't control the stock inside the buffer tank.
6. If the stock goes below the minimum:
   Produce in the red windows.
   Produce on back-up line 7.

7. If the stock goes above the maximum, wait until the next cycle to adjust.
8. Review parameters of the cycle and buffer two weeks before the next cycle.
9. Check R&P availability two days before production in the factory.
10. Check R&P contract for the whole cycle.
11. Trials will be done on the back-up line 7 or during the red windows.
12. Any change to the green-stream plan requires joint authorization by the production and supply chain directors.

As part of the rules, some companies use a visual representation to show what should be done if any items exceed the buffer limits. It shows inventory getting increasingly wide of the limits along with the appropriate action, depending on how far outside the limits it is. Figure 4.8 shows an example.

Two points are worth highlighting. Firstly, when only a small amount is outside of the limits, ±10% of the limit in this case, the agreed action is: "Do not change

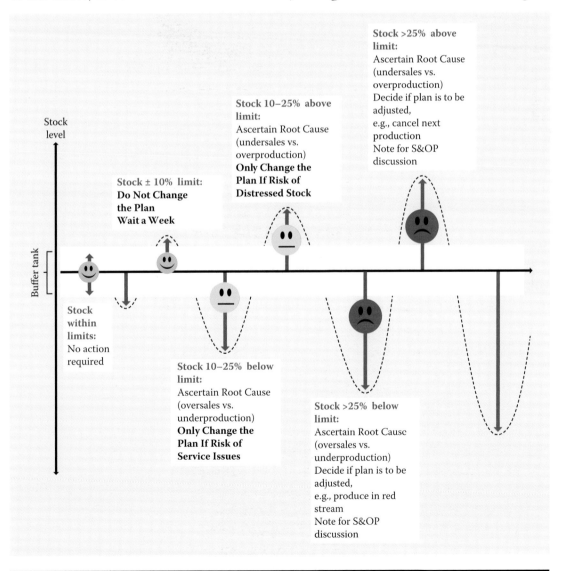

Figure 4.8 Stock alert chart.

the plan; wait a week." In other words, even when the limit is exceeded, it does not automatically trigger a change to the green-stream fixed plan. Experience shows that if one waits another week, there generally will be an opposite change in demand that will bring the inventory level back within the limits. Secondly, the actions are an escalation—rather than a single response—giving appropriate action, depending on how severely the inventory is outside the limits.

What about short-shelf-life products where one cannot hold stock? How can one apply the buffer tank principle to these types of products? It would appear impossible, but not so. A manufacturer of sandwiches supplying a major supermarket in the UK had a frenetic operation. They saw their demand as highly variable and unpredictable. The supermarket would give them advance orders midday each day of the week for the following day's deliveries. At 7 p.m., the supermarket would update this order based on the sales that day. The result was that this firm order was *always* different than the advance order. As the sandwiches had practically no shelf life and this company operated on small margins, any wasted sandwiches were a big issue. As they knew the firm order always changed, and each day's order was different, they believed that if they started to make sandwiches based on the advance order they would potentially make the wrong amount and throw sandwiches away. So the factory waited for the firm order at 7 p.m. before committing to supply and production. That was why the situation in their factory and suppliers became frenetic as they threw themselves into the task of producing exactly the right quantity of sandwiches ready to be delivered before 7 a.m. the next day to the retailer to be fresh on the shelf for their hungry customers.

A Glenday Sieve analysis was carried out on its products. As usual, although they had over twenty different types of sandwiches, just three accounted for over 50% of their volume. We asked the factory manager if he had ever looked at a graph of daily sales by sandwich type, in particular for the "green" biggest sellers. He hadn't, so we produced the graphs. Figure 4.9 is the graph of daily sales for the biggest seller. The other two big sellers had very similar patterns.

Sales in the first couple of weeks do look very variable. However, this covers the Christmas and New Year period, when one would expect sales to be unusual. After this, there is a recognizable weekly pattern. Sales vary day by day, but comparing sales for the same day of the week, each week shows a degree of consistency. Just like the hospital A&E department, the perception of variability was higher than the reality for the green items. The variability of sales for each specific day was relatively small and therefore predictable. This meant that the factory and its suppliers could start producing the green products much earlier than 7 p.m., when the firm order was received. Production volumes for these three sandwiches were planned to be a little less than the predicted volumes. Final small adjustments could then be made to the volume produced to match the firm order exactly when it was received. The company now had 50% of their volume made shortly after 7 p.m., and this meant more time to focus on the other, more variable, 50%. It would not work every week of the year, as Christmas had shown.

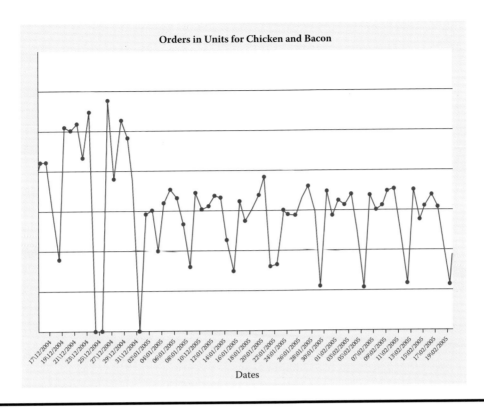

Figure 4.9   Sandwich sales variability.

However, it would work for at least 80% of the year. The weeks it would not work were well known, as they were when public holidays occurred.

The concept is called *repetitive flexibility*: Be repetitive and routine where you can to gain the benefits of economies of repetition. Understand where you can't, and then be flexible. This company could not have a buffer tank, but they could achieve the same effect, which was to protect a fixed plan for greens. The fixed plan was the predicted volumes prior to receiving the firm order at 7 p.m. The key was to have these production runs timed each day to finish just after the firm order time so that final adjustments to quantities could be made to match the firm order exactly.

Integrating Lean/RfS into Existing Processes and Systems

Some companies saw Lean/RfS as a pilot or initiative. We found this to be a big mistake. It was managed as something separate to the normal way of working, resulting in Lean/RfS becoming detached from it. This gave rise to some conflicts between Lean/RfS and the way of working in a batch environment. An initial implementation on one line may have been an experiment or pilot, but once a decision has been taken to adopt the Lean/RfS concept, it must be integrated into the normal way of working. Will this mean that some things will need to change? You bet. Moving from batch to flow logic is a paradigm shift. So it only stands to reason that some of the previous "normal ways of working" will need to change to adapt to Lean/RfS ways of working.

We have found it helpful to examine the existing supply chain planning and management process. What regular meetings and performance reviews happen now? What input data is required and what are the outputs? Who attends? There has usually been, at some time, a clear definition of what is meant to happen, who is supposed to attend, and what the agenda items are. However, it is not unusual to find that the disciplines around these regular planning and control meetings have become a little lax. For instance, attendees do not turn up or send deputies. Decisions are made without due reference to the data. Outputs are not clear. Lean/R*f*S is a good opportunity to refresh these disciplines and to make the process more robust. For each meeting and review, one needs to look at the inputs, the required outputs, who should attend, and how frequently the meeting should happen—then how these need to change due to the integration of Lean/R*f*S.

Mostly, the changes needed are relatively minor. For example, inventory targets become buffer limits. Most of the meetings stay at the same frequency and with the same attendees. Somewhere in the process the re-sieve needs to be incorporated. This is where, toward the end of a cycle, usually two weeks before it finishes, the next cycle needs to be agreed upon. This involves deciding which products are green, setting the production demands and buffer limits, determining the length of the cycle, and identifying any projects or similar factors that should be taken into account when agreeing on the next green-stream schedule.

The flow chart in Figure 4.10 shows how Lean/R*f*S was integrated into normal working in one company. It is not the "right" process. It is, however, something that one company developed and used. Similar versions have been developed in other companies we work with. Figure 4.10 is shown as an illustration to help people understand what is required.

Key performance indicators (KPIs) are another area that must be looked at to ensure that Lean/R*f*S is integrated into the normal ways of working. As the saying goes, "What gets measured gets done," so KPIs need to reflect what one wishes to monitor and track with Lean/R*f*S. If the KPIs in production and the supply chain are not aligned with Lean/R*f*S, then there will be confusion as to what people should be focusing on. In a batch-logic environment, it is often efficiency or OEE that is the number one focus for manufacturing. It is the first KPI that everyone in operations looks at. It is the prime focus of attention and scrutiny. In a flow environment, conformance to plan is the most important KPI. All functions in the supply chain should be focused on meeting the plan and delivering what is expected reliably, consistently, and at the quality required.

Removing variability of supply is a key objective of Lean/R*f*S. Variability of supply is a driver of higher inventories and excess fixed costs. The prime objective of the first steps of leveling—a fixed every-product every-cycle plan and the foundation of the Toyota Production System as well as Lean/R*f*S—is to create stability. From this, sustainable continuous improvement becomes much easier to achieve. This shift from efficiency and OEE to meeting the plan is often difficult for manufacturing people to do, especially senior ones, but it is crucial. The KPI used to measure conformance to plan is TIP/TOP (Total Item Performance/Total

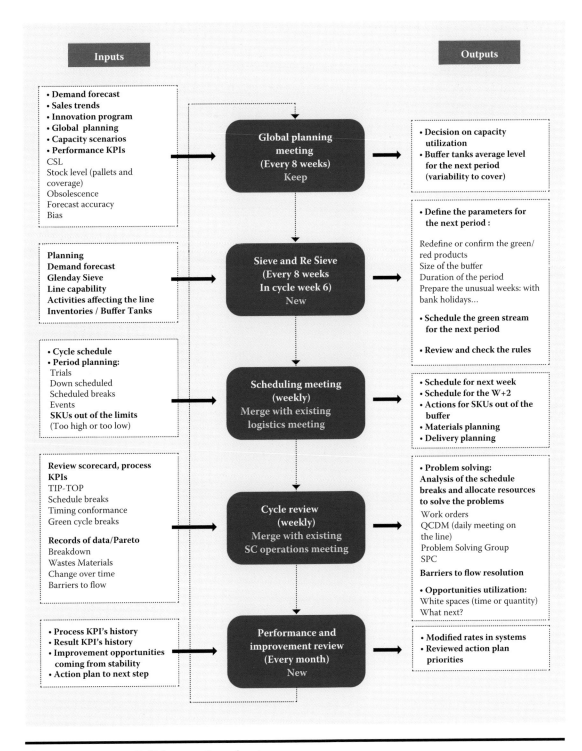

Figure 4.10 Lean/RfS integration chart.

Output Performance). Any over- or underproduction by item compared to plan counts as missed conformance to plan. For all items, total volume produced is expressed as a percent of total planned volume. With TOP, it is possible to have over 100% performance, which is not possible with TIP. How to calculate TIP/TOP and an example of the amazing results one can achieve are covered in more detail in Chapter 6.

What are the right KPIs and their targets? This is another area where there are no "right" answers. They should be agreed upon by discussion between all the functions in the supply chain. An example of what one company is using is shown in the Appendix, which shows their Lean/R∫S scorecard as well as definitions for each KPI. Note how they differentiate between "process" and "results" KPIs. The process KPIs are monitoring the Lean/R∫S process itself. Is the process working? If not, where is it not working? What do we need to investigate to understand the root causes of what has not worked? The results track the expected benefits or outputs. We often find that a company's existing KPIs only focus on the results rather than the processes that determine the results. We show this example as an aid to starting the debate on which KPIs are right for your organization to monitor the process, track the benefits, and help support Lean/R∫S. We are not saying that this example is what your company should use. As with the rules, the debate and agreement between the various functions is as important as what is agreed to be measured.

There are implications for the IT systems in integrating Lean/R∫S into the way the business operates. The existing planning systems will have been designed to support batch logic. It should be no surprise then that they are not programmed to operate to the concept of flow logic. With batch logic, the aim is for the planning system to determine the production plan, calculate the material requirements, and then record all the transactions that happen in the execution of these plans. With flow logic, the latter two activities are still required to happen within the planning system. The IT system should be fine as it is; however, there are two aspects to consider. With the materials requirement planning, one should ensure that any economic order quantities being used do not result in artificially created peaks and troughs of demand to the suppliers. Ideally, they would supply each week just what is required for the fixed quantities in the green stream. The other consideration is in recording all the transactions that happen in the execution. We have shown that batch logic requires perfect data; otherwise, the plan keeps changing—which of course it usually does because the data are not perfect!

A trend in supply chain management that has emerged in recent years is the belief that tracking and increasing the number of recorded steps will lead to greater data accuracy, even to the point of having radio-frequency identification tracking (RFID) for products to improve the efficiency of inventory management. We believe that the more steps one tries to control, the greater is the opportunity for error. Certainly it will create greater volumes of data and the need for bigger and bigger computers to deal with this data. During our work with various companies, it has been interesting to hear feedback from people on the impact of increasing the tracking of material movements. Here are a few quotes from people about a certain well-known IT system used by many companies—referred to here as "system XXX," as we cannot, for obvious reasons, name the software company concerned.

"System XXX has ossified our processes."
"It now takes us longer to record production of the product in system XXX than it takes to actually make it."
"It seems to me that feeding system XXX information has become more important to management than actually making product."

With flow, one aims to have the same activities being repeated at the same time every cycle—right across the supply chain, all with minimum stock—to create routine. People like routine, and routine helps them to maintain standards and reduce the number of tasks or actions required, for example, movement of physical materials, communication activities (always a source of potential misunderstanding and subsequent error), and transactions in the IT system. Up to an 82% reduction in tasks associated with planning and execution of the plan for green items has been achieved this way, resulting in less work, fewer errors, and less stress for people—and less detailed tracking of stock movements in the computer.

With flow logic, people working together in the re-sieve meeting determine the fixed-cycle plan—not the planning system. Their plan needs to be fed into the IT system. This can usually be done using "firm planned orders" so that the system cannot change the green-stream plan, although it will still produce a plan for the red items, as before. This is a manual override, but one that is only required once every cycle. The reality of how production scheduling is currently done in many factories is that the planner actually produces the schedule by some means other than the "official" planning application, for instance on spreadsheets the planners have developed themselves. This plan is then entered into the "official" system using firm planned orders. The only difference is that, with flow logic, the fixed plan will cover a longer time horizon—covering the whole cycle, usually eight weeks.

Also required are the two buffer-tank reports, as described previously in the section discussing buffer tanks. Some systems have reporting tools that can be used to produce these. If this is not possible, then they are relatively easy reports to produce in an automated spreadsheet—and most planners are very adept at developing these—using data downloaded from the main system. In some companies, especially where literally millions of dollars have been spent on implementing a companywide integrated IT system, it is argued that everything must be done within this integrated system, not on spreadsheets. If this is stated in your company, we suggest asking your chief financial officer (CFO) how many spreadsheets they have on their laptop. We think you'll find there's quite a few associated with the financial and budget planning of the company. If the financial planning for the organization is in reality being carried out on spreadsheets—with data downloaded from the main system—then why not the green stream? This is particularly true when the planning application of the main system is based on batch logic, which we have shown is fundamentally flawed because it requires perfect data—which is impossible to achieve.

Some Results

When making the change from a different plan each week to a fixed green-stream cycle every week, there is often an immediate improvement in performance. Why is this? Where we have been involved in implementing Lean/R*f*S, at KC and other businesses, we used a five-day focused workshop approach. The week creates a lot of excitement and energy. There are a series of teams working on different aspects of implementation. The best possible sequencing of products is agreed upon and owned by the operators. Rockbusting on the line identifies and resolves issues that are causing variable output rates. (Rockbusting is described in more detail in Chapter 6.) It can all add up to immediate benefits. Here are a few examples.

A Danish brewery's main products were cans of different beers supplied to discount warehouses. The contract with each discount warehouse was for a specific volume of each type of beer, so the brewery had to produce to a specific quantity; it could not run to time. The canning plant ran twenty-four hours a day, seven days a week, based on average efficiency, to meet the contract volumes. The company had sold the total volume they could produce in the week, and any overproduction would have resulted in cans of beer they could not sell during the existing contract. Therefore, the fixed green-stream rule was, when the required volume had been made, to stop the line. This would create "white space" time. The brewery management was asked to ensure that they had other activities organized for people to do when white space happened. One of these activities was 5S—something they had struggled to implement previously due to lack of time, as operators were fully engaged in keeping the lines running. The amount of white space due to higher efficiencies generated from the economies of repetition created each week after Lean/R*f*S was implemented as shown in Figure 4.11.

Every week after implementation, white space was created. By Week 6, twenty-seven hours of white space—a 16% improvement—was achieved. This was way beyond what the factory management thought possible. This was

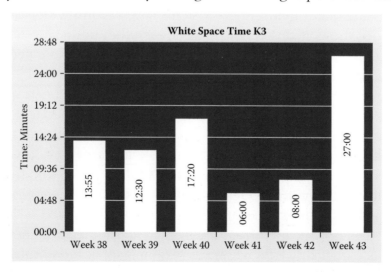

Figure 4.11 White-space generation.

extra potential capacity that could be sold to raise sales and profits when the contracts were renewed.

Another example centers on a biscuit company in Spain. Eight weeks after the implementation of Lean/RfS, the central planner had two stories to tell. The factory ran five days a week. Previously, she would plan only 50% of normal output for the first shift each week because of start-up losses and issues. This often resulted in overtime on Saturdays to catch up on production, which increased costs. In the first re-sieve—to plan the next eight-week cycle—she was able to plan Monday outputs the same as any other day, as the start-up issues had been eliminated. The plant had not run any overtime on Saturdays for weeks, as by halfway through the first cycle they were producing the same amount during the first shift as any other. This was accomplished by just starting and finishing the week on the same green products. Note that this plant was running to time, not quantity, so if they ran well, they produced more in the same time.

The second story she told concerned a product exported to the Middle East that was a biscuit made to a unique recipe and shape. Previously, the factory manager wanted to maximize the length of the production run for this product to minimize the large losses incurred by the changeovers to and from this unique biscuit. This typically resulted in a quantity equivalent to the customer of three to four months of inventory. The factory would then make it all in one go, which took about three and one-half days. It also meant that orders were only received every three to four months. The central planner had rung the factory manager to let him know the customer had placed an order for this product. To make it would mean stopping the green stream for a week. There was a period of silence on the phone as the factory manager thought about this. He then said, "No. Can we make it in three separate lots once a month in the red space, as I don't want to disrupt my green stream." The central planner was amazed to hear the factory manager asking for smaller runs with more changeovers. He would never have accepted that in the old ways of working with batch logic. She rang the customer, who was also delighted, as he got fresher product and lower inventories. The mindset had been changed.

Following the first implementation at KC, we returned to speak to people at the mill. We asked one shift manager if Lean/RfS had made any difference to him. His answer was, "Yes—a lot." He explained that, previously, the first thing he did when he came to work was to check the schedule, as usually it had changed in some way. He would then go into the material warehouse to physically check on whether they had enough of the right materials, and invariably they didn't. So he would ring the planner and together they would agree upon a plan that could be made as all the right materials became available. This process would take anything from one to three hours at the beginning of each shift. The shift manager said, "I haven't been in the warehouse for ages. I don't need to because I *know* the right materials will be there now. That means I've got time I never had before to work with my team on any issues they are having. And the results of that means the lines are running smoother and better." We have seen similar experiences time and again at factories. Not only does Lean/R∫S improve performance, but it also makes it less stressful and more productive, with greater motivation for people.

Summary of Chapter 4

It seems incredible that most planning systems are based on batch logic that is flawed—logic that is guaranteed to create a different plan every time. Why have more people not recognized this issue? We believe it's down to the power of conventional wisdom. If everyone is using the same logic, then surely it must be right. Economic order quantity is still taught in supply chain education as the right logic to use in planning systems for both production scheduling and ordering from suppliers. Challenging conventional wisdom is difficult. Anyone who does is usually seen as the one who is wrong and is ridiculed.

The good news is that more and more people *are* recognizing why batch logic causes firefighting. They *are* realizing their data will *never* be perfect. They understand there is another way: flow logic. The Lean/R∫S cornerstones are the crucial changes needed to make a paradigm shift from batch to flow.

Lean/R∫S delivers results that are very positive, far greater and much quicker than people imagined. However there is a downside. People see this as "having done Lean/R∫S." It has been implemented, and so performance in manufacturing has improved, more than they expected. But it is not seen as a foundation for further application across the whole supply chain and throughout all business functions. People often do not see the opportunity to create an integrated and comprehensive approach that will achieve a Lean enterprise focused on increasing profit margins, improving customer service, and boosting market share. The rest of this book is about how this opportunity can be exploited.

Chapter 5

The Lean/R*f*S Straight Edges

Let's continue with the analogy of our jigsaw puzzle. If the corner pieces provide the starting points, then the straight edges provide the shape and reference points to help build the rest of the puzzle.

There are three types of Lean/R*f*S straight edge pieces:

Firstly, the aspects that are not part of the paradigm shift of logic—the corner pieces—are now possible to achieve because of the logic change of batch to flow. These are activities that seemed practically impossible to achieve in a firefighting batch environment. Now, in a repeating green-stream pattern of production, they become possible—and desirable—to accomplish.

Secondly, the application of the principles of Lean/R*f*S across all functions in the organization makes it possible to use the Glenday Sieve in identifying green tasks and then to put these into repetitive, time-based, and standardized routines so that processes run better. It also means that people are less stressed, more motivated, and more effective in what they do.

Thirdly, and very importantly, are those aspects of Lean that have nothing to do with Lean/R*f*S concepts, but are crucial in making sure Lean/R*f*S is supported and sustained across the business. These tend to be related to Lean leadership, including how leaders in the organization direct and manage the way the business operates.

R*f*S-Dependent Straight Edges

Material—and Other—Flows

Just in time (JIT) is a concept that has lost popularity. It was talked and written about a lot back in the 1980s and 1990s, with many companies trying to implement JIT. The idea had come from Japan and seemed to make a lot of sense, that is, "Deliver just what is required when it is required" and "Keep inventories low and still be able to produce." The problem was that it was very difficult to achieve in a firefighting environment. Factories did not always have the right materials. Pressure was put onto suppliers to be "flexible," but often the only way they could respond to any demand was to hold more stock themselves. So their costs rose and so did their lead times—the opposite of what JIT was trying to achieve. Hence the practice was dropped by most companies that tried it. Nowadays it is a term everyone knows but few try to apply. It had gained popularity because some Japanese companies had been able to make it work, notably Toyota. What people did not recognize or understand was the fact that Toyota was not using batch logic to determine its production schedule. They were using leveled production—*heijunka*. The key difference was that batch created a different plan every week—and it then changed—whereas heijunka was a repeating stable plan. That meant that JIT was possible to achieve.

John Shook Insights into Heijunka

Back in 2001 when I first starting collaborating with Professor Dan Jones, he suggested I e-mail John Shook to discuss and compare the concepts of RfS with leveled production at Toyota. John was the first American to work as a regular employee of Toyota's global manufacturing headquarters in Toyota City. He was at Toyota for the beginning of the major internationalization phase of their development. The reason Dan suggested this was to get John's insight into how RfS fixed cycles compared with heijunka. Here are two extracts from John's emails.

Dear Ian and Dan,

First of all, Ian, I think you are right on in the direction of your thought. A couple of observations might be in order to illustrate how I think that is so.

Many meanings are contained in the term "heijunka." Unfortunately, translation often forces the translator to focus on just one aspect. You will hear Toyota sometimes refer to heijunka as the "objective" of system improvement, showing it at the culmination of a series of steps and evolution. Other times you will hear Toyota say heijunka is a "prerequisite" for attaining Lean that must be incorporated at the beginning. In fact, a "first stage" heijunka is something we sometimes call "pattern production." A pattern of running each part number repetitively, perhaps using an 80/20 rule to differentiate high from low runners, is established. This creates an "EPEC" that the organization can then strive to slowly reduce to smaller and smaller fractals. However, they eventually move toward "true heijunka," the primary difference of which is that now the customer becomes a part of the production process itself. This is a highly evolved production system that, as you correctly state, demands the integration of the entire enterprise.

John Shook

> I found it fascinating how early heijunka and RfS fixed cycles based on green items were in effect the same—developed separately, yet the same. Toyota's first-stage name said it all: "patterned production."
> A second e-mail from John had some more revealing insights.

Ian,

Your point that what is remarkable is that so few industries/companies/people have arrived at this understanding of heijunka is indeed interesting. I have stumbled across no one, until you, who had discovered this on their own. I think, as you suggest, more people *should* be able to understand these concepts.

I gave previously a brief, quickly crafted, maybe not 100% accurate, definition of the difference between heijunka and pattern production. Pattern production is simpler—repeat the determined pattern or cycle of production to create stability as a foundation. Heijunka brings the customer into the "system," creating a kind of customer–production system synthesis. And Toyota certainly *does* recognize the "economies of repetition" that you describe.

John

> Three further insights. Firstly is how so few understand that leveled production starts with a repeating pattern. Secondly is that the objective of this pattern is stability and not flexibility, which comes with the later steps of leveling. The last point really surprised me—the fact that Toyota recognized that patterned production created "economies of repetition." It is a phenomenon they knew occurred but had never given it a name.

When I first started at Kimberly-Clark in the mid-1980s, JIT for production materials was all the rage. There was interest in KC to do JIT and see the benefits of lower inventory. It didn't last very long, and it's not surprising. Since our production schedules were not stable, bringing in materials JIT wasn't feasible. You could "plan" to bring materials in right before the item was scheduled to be produced, but all it took was one small schedule mishap and the whole plan fell apart. So, the first time we shut down the operation for lack of materials was the end of JIT.

The amount of wasted time and effort to manage materials in a batch environment is staggering. It is not unusual for planners to do a daily check of all materials needed for the second and third shift each day. At KC, one of the production supervisors was notorious for calling planners at night and saying, "We can't find the materials for the upcoming production run. Where are they?" I remember getting these calls and telling him that I knew the materials are available since I checked on them before I left work. I could even remember the bay numbers where they were located. Talk about non-value-added wasted effort!

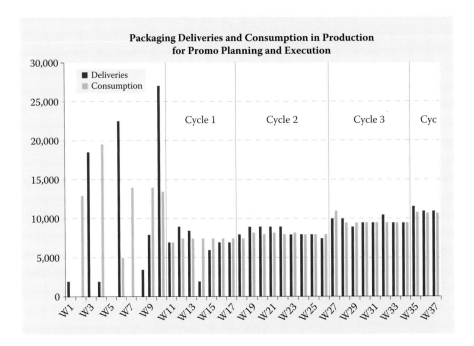

Figure 5.1 Packaging deliveries and consumption in production.

In a Lean/R∫S environment, the opportunity to apply JIT to materials for production is totally changed. The key is having a predictable *and* stable schedule. It should not be too difficult, once there is a fixed green-stream cycle running, to match supplier deliveries to this. There are several "watch outs." Ensure the material requirements planning (MRP) system does not create peaks and troughs in demand to suppliers due to preset order quantities. The suppliers should have visibility of exactly what the demand is. Check if the contract that purchasing has with the supplier includes a plus-or-minus allowance for delivery quantities to be different than the ordered quantities. How is "on-time delivery" currently monitored? Is it sufficiently accurate to support JIT? Through working together, you can adapt the deliveries so that exactly what is required gets delivered when it is needed. This may mean that the supplier has to hold some stock if their production process rate does not match that of the final manufacturer. It can also take some time to get the delivery quantities to match the production needs.

Figure 5.1 shows what happened at one company when it implemented Lean/R∫S. The blue bars show deliveries of a key packaging component from the supplier. The red bars are consumption of that component in the manufacture of the final product.

Prior to Lean/R∫S, both deliveries and consumption in production were variable, indeed erratic, with large amounts delivered and consumed followed by quite small amounts. Were these small amounts the result of firefighting? Probably. During Cycle 1, an attempt was made to match deliveries to consumption. However, after years of operating with batch logic, the supplier did not quite achieve it. The factory did manage pretty consistent consumption—the

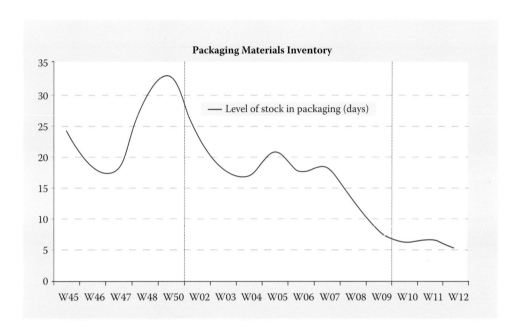

Figure 5.2 Impact on total packaging inventory level.

production volume of the finished product—in each of the eight weeks. Not bad for a first fixed green stream. In Cycle 2, it was decided to bring in a bit more of the packaging component than the demand in the first four weeks. This was for two reasons. The first was to make sure that a sufficient supply of the component was available for production. The second was to show that the supplier could deliver exactly the same quantity at the same time each week. By Week 5 of Cycle 2, the confidence had risen such that deliveries could be matched exactly to planned consumption. Now deliveries were no longer being put away in the material warehouse and then picked at a later stage for production. They were delivered directly to the shop floor—and that is just-in-time production. It also cut out a lot of non-value-adding material moves and data-entry tasks. Cycle 3 had an increase in production quantities, as white space had been created through economies of repetition, leading to extra capacity. One can see that it took a little time to synchronize deliveries to consumption as well as stabilize outputs at the fixed increased rate. However, this "instability" was minor compared to the previous situation before Lean/RfS.

What about the impact on the total inventory level for packaging materials in the warehouse? This is shown in Figure 5.2. It took just one and a half cycles to get the inventory down from an average of around twenty-five days for all components to just five days. It must be said that this company was very aggressive in wanting inventories down to release cash. It was willing to take some risks during the first cycles. Experience with other companies proves that they are more cautious and take a longer to achieve raw and packaging inventory reductions from JIT deliveries. But it can be done with Lean/RfS.

This example also illustrates the relative calm that is achieved with flow. In the situation seen before flow, what were the chances of mistakes being made? How much

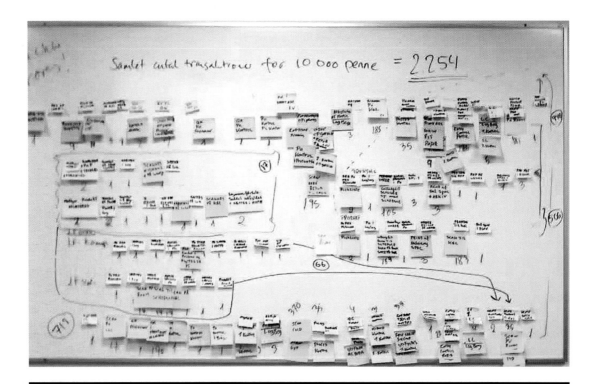

Figure 5.3   Example of task mapping.

time and effort was going into planning and managing this situation? With batch logic, it is like every day and every production run is a unique event that has to be planned, communicated, executed, recorded, and monitored. How many different activities are required to achieve this? It is an interesting question, one we wanted to find the answer to—to actually count how many activities and tasks were involved.

What was needed was a value-stream map, but not quite in the way it was usually done. The aim was to map in detail all the activities or tasks required to plan and physically get the materials to the production line for one batch of a green SKU. We were not mapping the actual manufacturing itself, but just the tasks required up to the point of being able to begin making the product, including all the computer inputs, phone calls, physical movement of stock, issuing of schedules, etc. We wanted to know about each and every time someone does something. We call these "touches." This mapping is best done by a team of people directly involved in the process. They use color-coded Post-It® notes to indicate the different types of tasks. A part of one such map is shown in Figure 5.3—it was too large to get into one photo! The number of tasks involved is always truly astounding to people. In this case, it was 2,254 separate tasks that were required every time they wanted to get all the materials available to be able to make just one batch of product.

The team looked at what tasks could be eliminated due to having a fixed cycle plan, i.e., what tasks could become routine and not need fresh instructions every time, what were duplicate activities, and so on. This team was able to reduce the total tasks required to 481. This is still a lot of activity, but it is an 81% reduction from the previous tasks. This really is eliminating non-value-adding

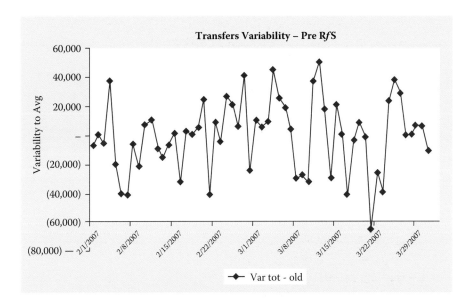

Figure 5.4 Actual distribution central to regional depots.

waste! Think of the time this can free up for people to work on making even more improvements. Think about how this would reduce the chances of errors being made.

This principle can be applied right across the supply chain using the "drumbeat" of production to create a regular predictable and stable flow of materials, information, and, ultimately, cash. At Kimberly-Clark, they looked at distribution from a factory central warehouse to a regional warehouse. Previously, when and how much to send was calculated using a distribution-planning software package based on batch logic. The resultant deliveries—rather lumpy—are shown in Figure 5.4. This is KC planning and delivering to itself.

Using the Lean/RfS principles of fixed cycles, a repeating weekly delivery pattern was worked out. This included higher amounts earlier in the week to correspond to the retailers wanting more product midweek in readiness for higher consumer sales on Fridays and Saturdays. This way, the inventory was planned to be as low as possible. This new way of operating needed far fewer activities to plan and manage. The new delivery pattern is shown in Figure 5.5 with a repeating weekly pattern of deliveries.

This same concept was applied at another company around shipments to customers. There were some differences, though. Firstly, they considered not green products, but green *trucks*. Not all the sales demand for each SKU was included on the green trucks. They fixed "green demand" at a volume that the customer was guaranteed to order each week based on previous sales history, as shown in Figure 5.6.

The second difference was that there were four green trucks; in other words, which SKUs and in what quantities differed between green trucks. Each week the customer got exactly the same four green trucks, as shown in Figure 5.7. The SKU product code and number of pallets (P) for each truck is shown.

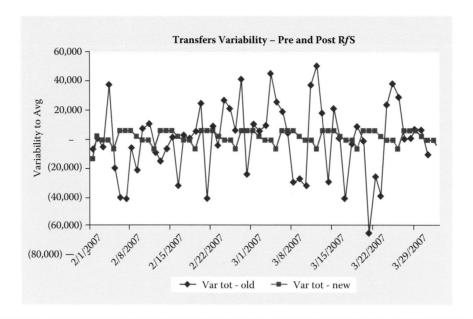

Figure 5.5   New delivery pattern with Lean/RfS.

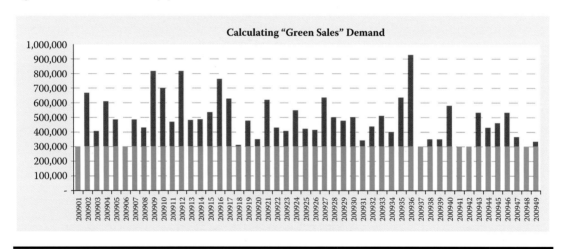

Figure 5.6   Total demand split into green and red demand.

KC Experience with "Shippers"

Sometimes the benefit of repetitive and routine material flow is not just in lower inventories and a reduction in the number of activities. An example of additional benefits came from KC and involved corrugated cardboard. In the United States, KC packs most of their products into cardboard cases, called "shippers," to protect the product during transit. It was not unusual to find the case packer being the bottleneck on the production line. These were fully automatic machines that erected, filled, and glued the cases. Any variability in the cases caused jams that stopped the line.

Using batch logic, KC would order big batches of cases—to get volume discounts—and keep some stock on hand, as they were never quite sure when planning would decide to run a particular SKU. Consequently, it was usually more than a week or even longer between when a case was made at the supplier and run on the line at KC. It was the normal way of working. What the case supplier and KC's technical people knew was that cardboard runs much better

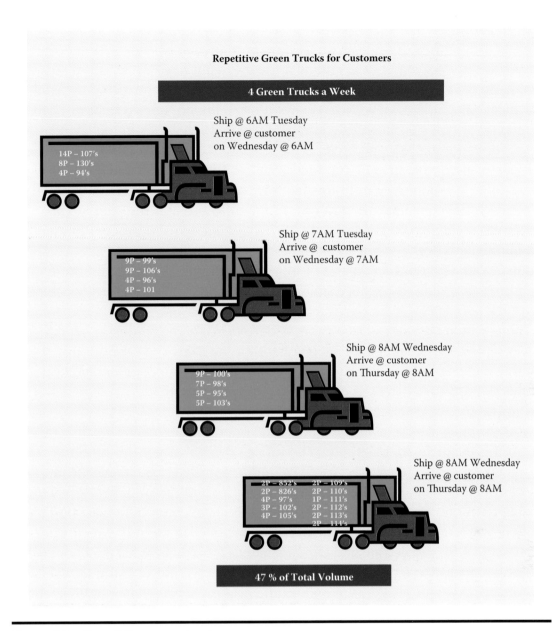

Figure 5.7   Repetitive green trucks.

when it is "fresh." Cardboard is susceptible to moisture and temperature changes that can warp it. The points where it has been cut or folded become less flexible, so that over time these folds do not work well when the flat cardboard is erected into a shipper by the case packer. Despite previous attempts at JIT to overcome these issues, it had resulted in many occasions when the right stock of cases was not available. The decision was made to stay with holding inventory and ordering in big batches, despite the impact this had on cardboard quality.

During the second implementation of Lean/RfS at KC, it was agreed to work with the supplier of shippers to see if JIT could be achieved with the repetitive cycle. The aim was to use the shippers at KC less than twenty-four hours after the supplier had made them—a challenging target, but one the team achieved. The result was that the case packer ran significantly better, with fewer jams and stops. This, in turn, resulted in happier and less-stressed operators. It was a major improvement that could be rolled out across all of KC's plants as they implemented Lean/RfS.

Blues and Reds

Lean/RfS is not just about determining a green-stream fixed cycle. It is also about eliminating non-value-adding waste. Creating "drumbeat" and routine material flow means that it is possible to get rid of a lot of unnecessary activities. There are two other types of potentially non-value-adding categories in the Glenday Sieve. These are the blues and reds.

Blues relate not to specific products, but are "blue" opportunities. This is where there is complexity in things like raw or packaging materials that add no value to the consumer yet add cost to manufacture and the supply chain. We find that companies generally do not have a recognized process to identify and eliminate blues. It is not something management focuses on. As a result, organizations often have a wealth of blue opportunities. Eliminating these would increase margins, reduce operational complexity, and make life a bit easier for people.

There should be a recognized process developed to enable this to happen. At its simplest, this could be an agenda item for the Sales and Operations monthly meeting, with a standardized form for presenting potential blue opportunities. The proposed actions can then be approved or rejected. A simple format used in KC for achieving this is shown in Figure 5.8.

This example also demonstrates how blue opportunities can arise. The products in this example are toilet rolls. The proposal is to harmonize the sheet count on similar products. This is because having different sheet counts results in a 5% loss of efficiency and output due to changeover and ramp-up issues when moving from one size to the other. How did this sheet-count difference originally happen? It was to improve margin by having less fiber in the roll, which was presumably easy to calculate and demonstrate as a "saving." What was not taken into account was the increased complexity in the factory, resulting in an overall loss of capacity equivalent to three times the original saving! It is known as the law of unforeseen consequences.

Would you know of any "blue" examples in your business? The list of examples is huge. Here are a few to illustrate the sorts of things to look for:

- Over thirty different grades of resin used in car paint formulations when five would work.
- Different pallet types for the same SKU to different customers, yet when customers were asked, they all agreed start to using the same pallet.
- Different bottle shapes for different liquid products were harmonized to the same shape but different colors for different products. Consumers did not notice—and if they had, would they have cared?
- A confectionery manufacturer distributing product to different European countries made every product type in both thirty- and forty-count cases. No country took both case sizes in any product. The company agreed to move to one case size for all countries. The case packer was a bottleneck on

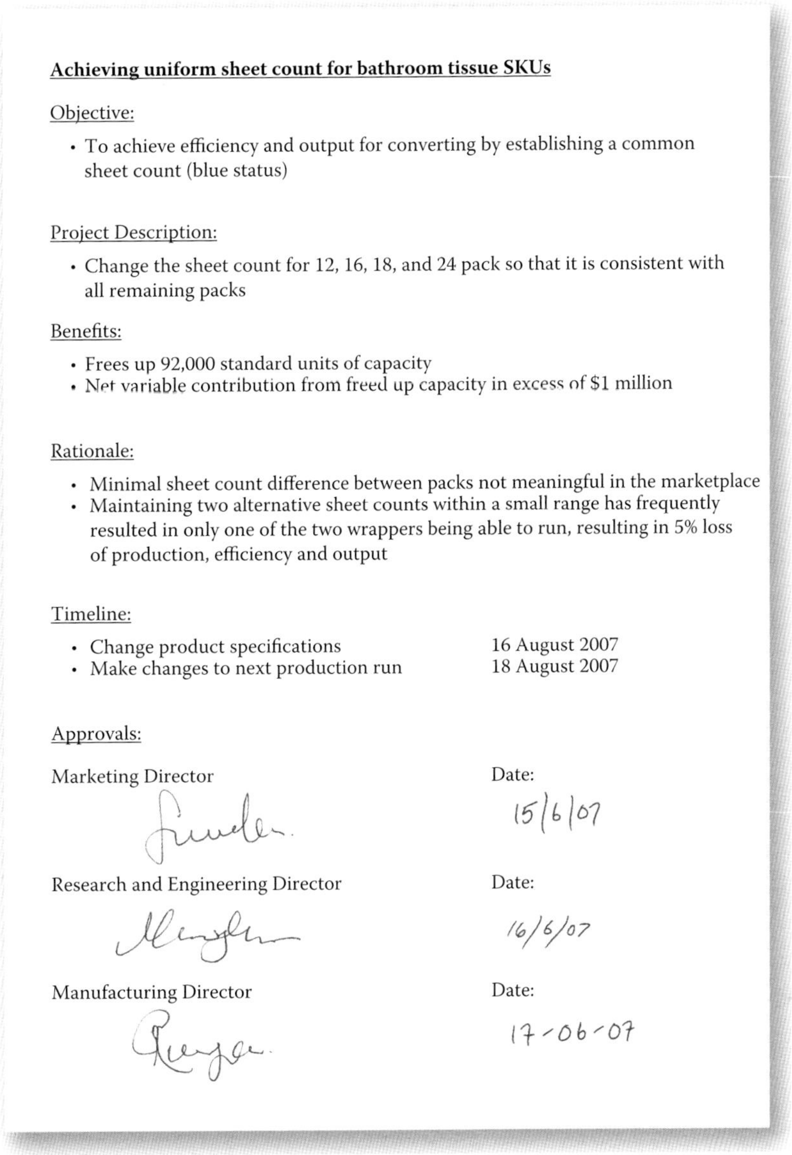

Achieving uniform sheet count for bathroom tissue SKUs

Objective:

- To achieve efficiency and output for converting by establishing a common sheet count (blue status)

Project Description:

- Change the sheet count for 12, 16, 18, and 24 pack so that it is consistent with all remaining packs

Benefits:

- Frees up 92,000 standard units of capacity
- Net variable contribution from freed up capacity in excess of $1 million

Rationale:

- Minimal sheet count difference between packs not meaningful in the marketplace
- Maintaining two alternative sheet counts within a small range has frequently resulted in only one of the two wrappers being able to run, resulting in 5% loss of production, efficiency and output

Timeline:

- Change product specifications   16 August 2007
- Make changes to next production run   18 August 2007

Approvals:

Marketing Director   Date:

     15/6/07

Research and Engineering Director   Date:

     16/6/07

Manufacturing Director   Date:

     17-06-07

Figure 5.8   Blue authorization form.

the production line, so the changeover saving increased line efficiency and product margins.

■ Date code in different position on pack for different products, meaning that the date code machine needed to be physically moved each time and reset to hit the date code space on the pack.

When looking at blue opportunities, it is not always possible to show a clear worthwhile saving for every proposal. This should not mean that only proposals that demonstrate savings should be approved for action. Removing unnecessary complexity that adds no value is an objective of Lean. The cumulative effect of many small reductions in complexity will add up to a better run supply chain and company—as well as happier and less-stressed-out employees.

We often hear it said, "Marketing will never agree to that." As a result of this perception, there is a risk that blue-opportunity proposals will not be suggested. Hopefully, the following example will show that it is definitely *not* the case that marketing automatically rejects blue opportunities.

The first Lean/R/S 5-day rapid implementation workshop at KC included a team looking for blue opportunities to implement during the week. Several were identified; however, one opportunity had by far the biggest potential, but it was seen as impossible to achieve. Why? People believed that marketing would never agree to it.

The factory produced Kleenex® facial tissues, including bundle packs—two or more boxes bound together by shrink film. At first, the quantities required were quite low compared to overall volumes, so the bundling was done as an offline operation. The shrink film used was clear, as the cost of printed film would have been considerably more. Thus a new label detailing the product with its barcode was attached to the outside of the bundle.

This proved to be a popular item with consumers, and demand from retailers started to go up quite a lot, but the offline bundling became too slow. An inline bundler and shrink wrapper was required. However, a problem had arisen with the current bundle pack at the checkout at the retailers. It was possible, when the product was being bought by a consumer, that the barcode on one of the inside single boxes was read instead of the new label due to the shrink-film wrap being clear. The consumer got a bundle pack for the price of a single box, and so was probably quite pleased. The retailer was less than pleased.

A technical solution was developed in the shape of a new machine called the "orientator." This was a large machine installed into the line that turned some packs around so that all the barcodes on the single packs were turned inside. The barcodes were not visible any more, as shown in Figure 5.9. The lower box

Figure 5.9   Bundle pack.

of tissues has been turned around relative to the other two boxes. The bundle pack continued to be a popular promotion with consumers. On normal production, the orientator was bypassed and so did not impact the output efficiency; however, when making bundle packs, the orientator was the bottleneck on the line, slowing down overall outputs and reducing efficiency.

The blue opportunity was to move to printed shrink film, thereby covering the barcodes on the single boxes and making them unreadable. Rapid-implementation workshops are about making changes happen in that week. The team wanted not only to raise the blue opportunity, but also to get agreement to implement it. A seemingly impossible task was made all the harder when they discovered that the marketing VP, who would need to ratify any decision, was on holiday with strict instructions not to be disturbed. The team decided to ignore his instructions and phoned him. Reproduced is the e-mail he sent to his marketing and sales team in response to the phone call he received.

Subject: Printed shrink film

Team,

As you may know, we have had some negative customer feedback regarding our bundle pack merchandising, specifically our ability to merchandise and brand on three sides of a bundle to enable pallet communication to be optimized. I recently sent out an email requesting that the organization look at ways to speed up getting to printed shrink film as one possible solution to the merchandising issue. Today I was informed that the operational leaders at New Milford would like to move immediately to printed shrink film on all bundle codes and then quickly accomplish the same at all producing sites. They didn't have to ask me twice if that was OK, I jumped in with both feet and committed our team to doing whatever is needed to support this accelerated changeover. We will need to make this a major priority going forward; graphics and specifications can be a bottleneck, so we are looking to find creative ways to move faster.

Additionally, several super premium bundles will need to move back to common case counts with their mainline bundle counter parts. Specifically, the 3-pack Ultra and Lotion upright bundles will want to move back to twelve bundles per case, common with mainline 3×100. And the 3×100 super premium flats will also want to be common case count with 3×200. I know that we had just moved down in case count on these codes with DIAMOND, but my understanding is that we were solving to common case counts with the single uprights on the Lotion and Ultra. I feel this is but a small change to go through to enable getting printed shrink film.

Please note that I'm on vacation, but this was a big enough "breakthrough" that I opted to come off vacation for fifteen minutes to type this note. Now it's back to 90 degrees, clear skies, pool, golf and several adult beverages.

Gary Keider

Clearly he is thrilled with the idea—not only to move to printed shrink film, but also to do it in an accelerated approach. Indeed, it would appear he was quite shocked at operations people wanting to move so fast! Things that could have been seen as obstacles that would prevent the change are seen as issues to be worked through and resolved. He directed his team that they should get on the case and get it done. (Although his last comment makes it clear—no more phone calls, I'm on holiday!)

The blue opportunity was agreed upon that week and implemented as soon as they had printed film available. The orientator was removed from the production line, and outputs on the line increased by 22% during production of the bundle packs.

Despite this success, it must be stated that we have found the blues process to be a very challenging process to sustain. Why is this? Firstly, the competitive nature of the retail environment continues to grow. Retailers look for ways to differentiate themselves from each other. They want unique product offerings from manufacturers, and this increases the number of SKUs. Secondly, the market has been complicated by the amount of innovation activity, both from a product-improvement perspective (where changes are made to improve performance) and in marketing activity (where graphics and pack designs are changed to increase consumer interest in the products). Eliminating blues helps reduce complexity. However, one must recognize that the market force to increase the SKU range and bring yet more complexity into the business is the current reality for most companies.

Reds are potentially another source of non-added value and complexity in the business. Low-volume products can create problems in planning, manufacturing, and matching up with the supply chain. Their demand tends to be highly variable at a percent level, making them unpredictable. There are usually lots of reasons put forward as to why these reds need to be included in the portfolio, including that the business should be responsive and flexible to meet demand for all products in the range, including the reds. If manufacturing has the physical ability to make them, then why shouldn't we make whatever the consumer wants? Just like Toyota.

Except Toyota doesn't make every possible car they could make—far from it. This was brought home by an experience at a pharmaceutical company that wanted some help to implement Lean/RfS. They had a well-developed Lean program already running supported by internal consultants. They had also been using a retired VP from a very large car manufacturer who was doing some consultancy work. He had been advising the pharmaceutical company that they needed to be able to do "one-piece flow matched to market pull through TAKT time"—the final step of leveling. However, he had not been advising them to implement the initial steps of patterned production, apparently because he was unaware of them. He said that this approach was what his previous company was attempting to do. Just like Toyota.

I asked him if anyone at this previous company had ever calculated how many potential SKUs they had, taking into account all the possible options available when ordering a new car. He replied, "Funnily enough, we did do that calculation." I asked, "And the answer was?" He replied, "Three million, give or take a few thousand." Absolutely amazing! This car company was trying to build a supply chain that could manufacture and deliver any one of 3 million SKUs just in time to the customer's demand. I asked him how many potential SKUs he thought Toyota had. "Probably about the same" was his response. Wrong! When customers look at the Toyota website to buy a new car, they are invited to design which specific car they want. There are many different options on type of wheels, seating, sound system, engine size, body color—all the usual sort of choices someone buying a car expects. The options appear endless, except they're not. Indeed, many of the possible combinations are not allowed. Many "options" are mandatory. For instance, choose an economy-sized engine and it is likely that you cannot also have leather seats, a wooden steering wheel, and fancy alloy wheels. Likewise, select a V6 3-liter sporty engine, and chances are you *have to* have the leather seats, wooden steering wheel, and fancy alloy wheels!

Toyota has packaged the options so only certain combinations are allowed, meaning that they have significantly fewer SKUs than it would at first appear. They do not make exactly what the customer may want—even if they have the ability to make all the potential combinations. It would prove just too complex and not cost effective. With most car manufacturers, it is possible to order a yellow-painted car. Not with Toyota (at least in Europe). They do not make yellow cars, as the number of customers wanting a yellow one is just too small to bother with and be able to meet demand profitably. Most manufacturing companies have around 30% of their SKUs accounting for the last 1% of sales, i.e., the reds. Is this really sensible? Do you really want to be like Toyota? If so, stop making the reds.

Which Cats Prefer Parsley?

A well-known pet-food company wanted to implement Lean/RfS on one of their major manufacturing lines. It produced cat food. The Glenday Sieve analysis of the range came out as expected, where 6% of SKUs accounted for 50% of the volume sold. Similarly, just 1% of sales equaled 30% of the range, a typical result. They had 220 different SKUs—for cats. The reason for this was quickly given: The factory supplied all the countries in Europe, so each country needed its own language label, hence all the different SKUs. This was to some extent understandable, even though multilanguage labeling is used on many human products.

So regarding cat food, one would have thought it was possible to reduce the range complexity. "How many different recipes are there?" was the next question. "Sixty-nine" was the answer. Sixty-nine separate cat food recipes! This seemed ridiculous. Why so many? It turned out that each country in Europe was a separate commercial division of the business. Each could decide what "their" cat food recipes would be. There was little coordination between the countries about recipes. There were three unique recipes for cats in Finland! How many cats are there in Finland, and what makes them different from the cats in the rest of Europe?

Speaking with the planner about the issues this number of recipes and SKUs created revealed that the minimum batch size possible, due to constraints in the processing equipment, was 150 kg. For the smaller-volume recipes, that gave over six months of inventory! The manufacturing plant had been designed and built on the premise that there would only be a limited range of high-volume standard recipe products sold across all European countries. But now they were making sixty-nine recipes and 220 SKUs on equipment not designed for this complexity of range.

We see this time and time again in companies. *The range grows*. It is produced on machinery that was not designed for small-volume products, yet the product costing system shows these items still make money. There was one last input about the issues of the range from the line operators. They were resigned to the number of recipes and the increase in the amount of clean-outs this caused—it is the way things are nowadays in companies. However, they did find one particular clean-out issue. One of the cat food recipes had fresh parsley in it that took a lot of extra effort for the operators to get all the little green bits of parsley out of the equipment. One operator observed, "I want to know which cat said it preferred parsley in its food and caused us all this extra work as a result!"

Red to Green—or Rather White to Pink!

At Colman's of Norwich in the UK, a main product category was Robinsons' squash. The catchphrase in Colman's marketing department for this important brand was "bought by Mum for kids." In other words, marketing had to convince "Mums" to buy it, but it was the "kids" they had to appeal to. It was a big brand with many SKUs, including different fruit flavors, bottle sizes, and case counts.

The Glenday Sieve analysis gave the usual results—30% of SKUs gave only 1% of sales. When the marketing director saw the sieve results, the decision was to de-list all the red SKUs. These included all the grapefruit-flavored products. These products had the same recipe based on white grapefruit, which had quite a sour taste. The different SKUs were due to many bottle sizes and case counts. The reasoning to de-list all grapefruit SKUs was that as Robinsons' squash was "bought by Mums for kids," it must be some pretty weird kids that preferred grapefruit.

With no grapefruit, Robinsons would switch to other fruit flavors and no sales would be lost. The de-listing of all the grapefruit SKUs resulted in the biggest number of customer complaints Colman's had every received for a single issue. Hundreds of letters arrived, and not one of them was written by a child. They were mostly written by people who had retired. As one marketing person put it, "Every single person who bought grapefruit must have written to us." It turned out that they enjoyed it because it was a drink that had a bit of "bite" to its flavor. It wasn't sweet or carbonated like most soft drinks. The marketing director was a very clever person and saw the potential, and it wasn't to just go back to the old small volume "red" SKUs.

Product development was given the task to revise the recipe based not on white grapefruit, but on the much more in vogue pink grapefruit. Instead of the previous variety of bottle and case sizes, there would be just one: 750 ml in a 12-count case. It was also decided to use the letters of complaint in a novel marketing campaign. Adverts were placed in magazines frequently bought by retirees. These adverts showed some of the letters to demonstrate the passion the people had about losing their grapefruit Squash. Instead of trying to suppress or explain away the level of complaints, the sheer number and feelings expressed was to be positively headlined in the adverts. The salespeople went to the retailers to explain the marketing campaign and get listings for the relaunch. The idea caught the imagination of many of the retailer buyers, as it was a new angle and a different approach in the product category. It was aimed not at kids but adults. Many agreed to special displays to help the relaunch. Sales exceeded all expectations. Pink grapefruit jumped from being a number of low-volume red SKUs to the third biggest selling SKU in the Robinsons' squash range: red to green—or rather white to pink—a trick executed by a very clever marketing director.

One of the problems with "red" products is that the current product costing system will often show that these items are profitable. Therefore, it's worth making and selling them, even if, for many people in the organization, they are just a "pain in the butt!" The reason this happens is, firstly, that total fixed costs are divided by total units sold—each unit gets the same allocation of fixed costs. Yet most people believe green items are easier to manage than red ones—they need less effort to plan, make, and sell. Secondly, when green and red items are made on the same equipment, the average efficiency of the line is used in the costing. No account is taken that they might actually have different efficiencies or that changeover times are often disproportionately higher for red SKUs. The greens are effectively subsidizing the reds, but the current product costing does not show this.

What is needed is a way to more accurately show just what the reds really cost.

Lean/RfS Product Costing

At one company that had implemented green and red streams, the senior managers complained that the improvements in output rates on the production line had not been as good as they had hoped for. The factory ran both green and red products on the same line—normal practice for many factories. They had seen outputs go up considerably on the days when they ran green SKUs; however, efficiencies had gone down on the days when they ran the red codes. Overall efficiencies had got up, but not by as much as they had hoped for.

This should be no surprise. What they were seeing was what they already subconsciously knew. The higher-volume products tended to run better than the lower-volume ones. The settings for optimum running are better understood and agreed upon by people for the frequently run high-volume items than the infrequently run low-volume ones. Also, time lost to ramp-up and cleaning was proportionally greater with the smaller items. However, the way most companies measure efficiency is an average across the week, indeed across several weeks, to get a demonstrated efficiency figure—and that is what is used in product costing. Effectively, the greens are subsidizing the reds. With batch logic, it may be recognized, but nothing can be done about it, as one never really knows when a particular product will be made. Each week is a different plan with the greens and reds mixed up. Separating the differences in efficiencies is very difficult to achieve.

With fixed green and red streams, one does know when they will be produced. This means that efficiency for each can be measured separately. This can then be used as a basis for costing the product. At first, accountants tend not to be too enthusiastic, as they want to be sure that it is a sustainable situation. It is something they will only adopt when manufacturing can demonstrate the discipline in sticking to the fixed green plan with fixed red days. It also makes collecting the data a little more involved for production. However, once people in manufacturing realize the potential significance of the change—getting rid of some reds—they are keen to start.

It is interesting to note that a second aspect starts to become apparent once green and red runs are measured separately. Invariably, people see that the number of stoppages and breakdowns on red days is greater than green days. Why? When asked, the response people give is that it is most likely because the settings of the line are not so well understood for the reds. This is because they are run less frequently and for short run lengths. As a result, more issues like jams and stops for adjustments happen when they are run.

Using the different demonstrated efficiency rates for greens versus reds means that their product costing is more realistic and accurate. The greens will be more profitable and the reds less so. That is not to say that the reds become unprofitable; they may still make money. However, the more accurate product costing will help people in the business make better decisions regarding how they manage and promote the products in the marketplace, i.e., whether the reds really are worthwhile to have as part of the product portfolio.

Of course, if one is able to produce green and red products on completely separate equipment, then current product costing will automatically include any differences in efficiency between them.

Applying RfS Principles Across the Business

As R*f*S was applied in manufacturing and the supply chain, it became apparent that this could work on any process. We took a look at several different business processes, for example, closing the monthly accounts, recruiting new people, performing promotional planning and execution, and implementing a price increase in sales. In each case, the process seemed to have the same issues of firefighting, variability, and lack of standards. We found that the principle of the Glenday Sieve—separating items into green and red categories—can be applied successfully in any function or to any business process. The difference is that, instead of looking at products, one is looking at tasks or activities. It makes the Glenday Sieve a more subjective than mathematical exercise, but it still can be done. One then creates fixed repetitive and routine schedules for the greens while being flexible when dealing with the reds. People become more efficient—achieving more for the same or less effort—and, importantly, more effective as well. Being effective means being focused on doing the *right* things.

When looking at business processes or functional practices, green items are those activities that are considered to have the most added value. So putting them into fixed repetitive routines helps to ensure that they get done and done well in a standard way. The reds tend to be the things people see as limited or non-added value that just consume their time. Think of e-mails, meetings, and unwelcome interruptions.

There is a difficulty. It is relatively easy to analyze products using the Glenday Sieve. One can use actual or forecasted volumes or values. Whatever data is used, it is easy to calculate the percentages and then apply people's opinions to finalize the choice of which products are green and which are red. Application of the Glenday Sieve analysis tool to business processes or practices within functions is more subjective, as usually there is no mathematical data available. However, it is still possible to do it and obtain the benefits of economies of repetition. Determining which activities or tasks create the most added value needs to be decided in discussion and debate. Doing this greatly helps clarify what people should be focusing on and which activities they need to treat as "green"—the few (6%) tasks that provide the most added value and thus should be standardized, disciplined, and routinely achieved in a "drumbeat" pattern to ensure that they get done.

Engineering Project Management

Here is an example from Kimberly-Clark to illustrate the approach—engineering project management.

Over the years, there had been several attempts to improve engineering project management. Consultants had been hired to develop a "world-class project management process," a complete process that was disciplined and standardized and that could be used for all projects. KC's definition of a project was anything and everything from buying and installing a relatively small piece of equipment up to building a new factory on a greenfield site. Many projects typically had issues of being late and over budget—not all but many—and hence the desire to have a "world-class project management process."

With Lean/RfS, we would agree with the desire but not with what had been tried. What had been attempted was to develop a single totally standardized end-to-end process for all projects. People are not robots, and they do not like to feel hemmed in by too many standards or being told what to do in every situation. They agree with the need for *some* standards; however, they also like to have areas where their own views and expertise can be allowed to determine the best approach or way forward. The name—Repetitive *flexible*—describes what is required. Where should one be repetitive, standardized, and routine (i.e., green) versus where to be flexible, allowing personal interpretation and experience to prevail (i.e., red)? The trick is to know which tasks or activities are which!

A group of engineers, managers, and quality personnel were brought together to work on a Lean/R*f*S solution to improve project management. They started with the process map produced by the previous set of consultants. Their complete map of the "world class" project management process contained over 500 separate activities. It was highly unlikely that anyone was ever going to correctly and faithfully follow over 500 steps. It was doomed to failure. Adapting the Glenday Sieve principle, the participants were set the task of picking just 6% of the steps in order to identify those steps that, if always done to a standard with rigor and discipline, would best address the issues of poor project management. It was a lengthy but necessary debate. It was necessary to ensure that all the participants understood why those particular tasks had been selected. To everyone's surprise, the 6% of tasks were all from three key areas of project management. These were:

- Agreement of the initial project proposal (scope, deliverables, time scale, etc.)
- Detailed costing and budget authorization
- Postproject review and lessons for future improvement of the project management process

The first two were all about setting the project up clearly and with as little ambiguity as possible. Historically, this often led to issues later in execution of projects, with people having different expectations, or misunderstandings, due to vagueness in how the project had initially been defined. There were disagreements of exactly what had, and had not, been included in the project budget. The last area—which everyone admitted was never really done—was about becoming a learning organization, identifying what issues had occurred during implementation of a project, and then using these to improve the project engineering process. These three areas were also the areas the engineers did not really enjoy doing. They much preferred actually doing the projects. A structured, disciplined approach would help them complete these key important green tasks to a higher standard.

Deciding on a methodology for these tasks was the next step. After some discussion, one engineer suggested that, rather than trying to define a methodology for these tasks themselves, they should use the methodology from a company that did it really well. The company he had in mind was a specialist engineering contractor that KC sometimes used. As project engineering was their sole business, they had developed very good systems to ensure that the project was well defined and properly budgeted with their clients before any project started. Then they analyzed how the project had proceeded to learn from any issues that had occurred so that they were not repeated. Their business depended on getting these areas right. As a consequence, their procedures, paperwork, and systems were pretty good, easy to follow, and, most importantly, worked. These were adopted by KC.

Everyone was really surprised that none of the activities involved in the actual process of managing and implementing the project itself were considered "green." It was interesting to ask the people involved why they thought this was. There were essentially three reasons. Firstly, each project was different. Some were big, some small, some complex or relatively simple, some repeats of similar projects from before, and some completely novel. To have one process with standard tasks covering all these different needs did not really fit any that well. The nature of the project determined the degree of sophistication required in the project management activities of the physical implementation aspects of a project. Secondly, this is what the engineers had been trained in, liked doing, and were actually quite good at. Lastly, there were various software tools available to use. Each engineer had their favorites, and these often depended on how sophisticated they needed to be, depending on each project. So it really did not matter what people used as long as the project was implemented according to the original agreed-upon proposal and costing, with any issues providing learning to be incorporated into improving the overall process.

The Lean/RfS principles, when applied to business processes, are as follows:

■ Do not attempt to define and standardize the whole process. People just won't follow a totally defined process.
■ Identify those 6% of the tasks or activities that most influence a successfully operating process.
■ Develop standardized, repetitive and, ideally, time-based procedures for these green tasks.

Lastly, implement a visual monitoring system whenever possible. This last point is illustrated in the next example.

Promotional planning and execution is one process in any consumer-branded business that always seems to have problems. There are many reasons for this, not least of which is that marketing personnel change frequently. Plus, the very skills and personality that marketing people are hired for—creativity, innovation, new ideas generation, and excitement—do not readily lend themselves to following standardized, disciplined processes. Promotional planning and execution is a process that has been improved in several companies using the Lean/RfS approach.

Promotional Planning and Execution

Here is one example from a well-known confectionery company. The approach was the same as before, to get a mix of people involved in the process together to jointly agree upon a solution. This group identified three different types of promotion and what percent each accounted for in the overall number of promotions:

■ Strategy and new technology = 10%
■ Existing product into a new market = 15%
■ Typical repeatable existing product into an existing market (extra free, on-pack flash, etc.) = 75%

It was the last type that needed a process that worked better. They brainstormed what currently was not working. The main issues were:

- Activities scope changed
- Too long to approve legal text
- Item code issued too late
- Promotional forecasting inaccurate
- Too long to collect input data
- Packing material, especially printed, not agreed upon within supplier lead times
- Difficulty of getting the listings of standard SKUs changed to promotional SKUs at the retailer's end
- Changes in activity phasing

Sound familiar to anyone?

They also brainstormed what was currently working well. There was only one thing they could think of: *firefighting*. A discussion took place on what was missing: What were the key tasks needed to make the most frequent type of promotion work well? They determined three tasks:

- Everything in place four weeks before promotion
- Rules and communications
- Monitoring of promotional sales

They agreed to the following Lean/R*f*S green principles:

- Process to be visually monitored and communicated
- Key tasks (greens) to be standardized
- Clear responsibility for each step, and product brand manager responsible for success of overall promotion
- Transparent timing for each promotion via visual management board
- Ability to learn from mistakes

The group developed a visual management board that was put up in the marketing department (see Table 5.1). It does not matter what this group chose as its green tasks. What is important is the discussion to agree upon which tasks are green so that everyone understands what they are and why.

At the onset of each new promotion, the green tasks were added to the board, and the date required for each green task was entered in the appropriate box. If the date was achieved, a green smiley face was put in the box. If it was not met, then a red sad face was added. This led to a big debate. Surely all the dates must be met to have a successful process? Not so, as what is required is a better process, one that works better most of the time. It is too much to expect it to work perfectly all the time. After all, this is promotional planning and execution, and things will go wrong at times—it is inevitable. What is needed is an awareness of when things have gone wrong, why it happened, and what can be done to learn from the experience to further improve the process, in order to become an organization that improves by learning from its mistakes. A slogan was added to the board:

It's not BAD to be red.

It's BAD not to know WHY!

Table 5.1

Visual management board

	Step	Market concept agreed (after pre-reading)	Draft P&L / APPACS ready	Activity Approval (P&L, Time, AW, SC)	POR Start	Final Market Input	Final Forecast	Final P&L (Activity approved)	Packs Order	First production
			improvement		improvement	improvement		improvement		improvement
	Responsible	Brand Manager	Brand Manager	Brand Manager	Site Activity Coordinator	Trade Marketing Coordinator	Demand Planner	Brand Manager	Packaging Coordinator	Production Planner
	Length	START	after 1-3 wks	after 1 wk	0 wks - execute	after 3 wks	after 1 wk	after 1 wk	5 wks to execute	2 wks to execute
No	Activity Name / Date	16-18	16-18	15	14	14	11	10	9	4
1	Candy bar shipper activation P08-P09	P03W4	P03W4	P04W1	P04W1	P04W2	P05W1	P05W2	P05W3	P06W4
2	Multibrand P10-P11	P05W4	P05W4	P06W1	P06W1	P06W2	P07W1	P07W2	P07W3	P08W4
3	Chocolate/Choc Delight P11-P12	P06W4	P06W4	P07W1	P07W1	P07W2	P08W1	P08W2	P08W3	P09W4

Certainly at first there was some resistance to having a visual management board up in the marketing department monitoring their performance. It had not been done before. Performance boards were something on the shop floor, not in the marketing office. Why not? Having visual management boards is a good way of helping improve performance—and it worked in this example of promotional planning and execution.

Why does visual management help? Apart from clarifying responsibilities and identifying expectations of what is supposed to happen when, there is another aspect that helps improve performance. It is called the Hawthorne effect, named after some experiments carried out in the 1920s at the Hawthorne works (a Western Electric factory near Chicago that made radios). It was believed that if just the right environmental conditions—light, temperature, air flow, etc.—could be achieved in the factory, then productivity would go up.

A series of experiments was carried out in a control room. Every time any environmental condition was changed, productivity went up. However, when the same condition was replicated in the main factory, nothing happened. After several experiments, it was recognized what was happening. The output of production in the control room was being monitored more closely—hour by hour—using visual output boards next to the operators. They could see more clearly what their performance was. Performance went up. It had nothing to do with the environmental conditions.

The same thing nearly always happens when one introduces easily understood visual monitoring boards of performance in any workplace. It is one of the reasons why Toyota has so many, usually hand updated and color-coded, visual display boards—and not just on the shop floor. No marketing person wanted to have red sad faces against their promotions. The whole process became better; people learned from any mistakes; and it was easy to explain to new people joining the department.

Lean/R/S principles can be applied to any business process to make it more effective. It has even been applied to how people manage their time. At Colman's of Norwich, the senior managers of the business felt that they were not being as effective as they could be. So the thought was, "How could the principles of green and red be applied to what we do each day?" The idea was fine; however, the difficulty was deciding which tasks were green and which red because, of course, all the tasks of a senior manager were important! An exercise to help decide was developed. It consisted of three brainstorms, one after another. The first was to brainstorm:

■ What should a good manager do?

This resulted in lots of responses and suggestions. The second topic was:

■ What does a coach or mentor do that a good manager does not do?

Any response people felt should have been on the first brainstorm was added to that sheet. Rarely did this brainstorm come up with any suggestions that people did not feel should go on the first list. The third topic was:

■ What do you spend most of your time actually doing?

The participants were then asked to cross out any of their responses that appeared on brainstorms 1 and 3. What was now left on list 1? It tended to be all the coaching/mentoring-type tasks that they wanted to put on list 2 yet already had on list 1. The things that usually appeared on both lists 1 and 3 were the more mundane, administrative, and even negative aspects of their jobs. What was missing from list 3 were the motivational, encouraging, and developmental aspects that they knew they should do but somehow never quite managed to achieve on a regular basis. Not everyone shared this issue. Some people were naturally very good at making sure they did these positive tasks in their daily routines. But they were the exception. The list of activities remaining on list 1 were the things that helped develop, motivate, and recognize people's achievements—and these were the green tasks for senior management. Each person was asked to develop their own weekly schedule for these green activities to ensure that they did them. Green time slots were coordinated for joint green activities. The rules were as follows:

– Activities in green time must happen except in cases of true emergency.
– No one is allowed to interrupt anyone during that person's green time.
– Phone calls, responding to or writing e-mails, and ad hoc meetings can only happen in red time.

Each person posted their weekly schedule in their office area for all to see. Some found this difficult to do. They were not used to operating to a schedule—even though it was only for a portion of their time. However everyone went along with it at Colman's – greatly helped by the managing director at the time being a big fan of the idea and an enthusiastic participant. It transformed the culture in the business. The management style changed from one that tended to be hierarchical and directive to one that was more empowering, led by example and with higher positive recognition of people's achievements. This in turn resulted in a surge in performance at all levels. It is the story told in Chapter 3 of this workbook, "Squash Quosh," that led to a doubling of sales per employee and market share plus an increase in net margin from 12.8% to 17.2%.

Toyota has a similar process. It's called *leader standard work*. Specific time slots in the day are allocated for certain tasks with free time from all other tasks.

It is part of creating Lean leadership in the organization, something KC found was a key piece of the jigsaw puzzle when it came to achieving a Lean transformation.

Other Straight Edges Supporting Lean/RfS in the Business

At Kimberly-Clark, implementing Lean/RfS in manufacturing and the logistics areas was seen as a success. Performance improved; people were happier; and targets were exceeded. However, there was a problem.

These improvements helped manufacturing and supply chain operations of the business to get better, which was good. However, these improvements were not really being used to lever increased competitiveness and drive, to beat the key competitors on all aspects—quality, price, market share, and margins—because two things were missing. The first was a clear strategy cascading throughout the business that focused everyone toward achieving that strategy. KC needed something that would help them harness and apply Lean/RfS across the whole business in a way that delivered increased marketplace success. Secondly, improved Lean leadership was required, meaning a different type of approach and behavior from management. Kimberly-Clark needed to become a better "learning organization," a business that focused on process improvement and problem solving rather than results and firefighting.

Policy or Strategy Deployment?

What is the difference between policy deployment and strategy deployment? They are really the same thing: The terms *policy* and *strategy* can be used interchangeably. Since Ian was taught it as policy deployment, he tends to call it that. At KC, we were taught by Pascal Dennis, and he tends to use the term strategy deployment—hence the use of both terms in this workbook.

Over a period of more than two years, Kimberly-Clark worked with Ian to implement Lean/RfS across many of our plants in several regions of the world. RfS cycles became the norm. We also applied green-stream and red-stream thinking to some other business processes. Lean/RfS gave us an understanding of how to stop firefighting and establish stability. The Rapid Improvement Workshop approach, which was linked to creating economies of repetition, engaged our people.

Fundamentally, it taught us to see and believe that we could "flow." We needed Lean/RfS to believe that stopping the firefighting was possible and would lead to the creation of a stable foundation for sustainable improvement.

However, Lean/RfS by itself has limitations. As time went on, team members and senior management changed roles. We started asking where we were going with Lean/RfS. We had demonstrated excellent results, but we needed to go further if we were going to embed this thinking and demonstrate clear marketplace improvements for our company. Without policy deployment and Lean leadership, we lacked the clear, aligned direction and deep thinking to both sustain and advance Lean/RfS to drive a true Lean transformation. This is what Pascal Dennis brought to KC. He taught us that if we wanted to be one of what he called the "5% club" of companies that successfully transformed to a Lean culture and breakthrough performance, we would need to add strategy deployment and effective Lean leadership pieces to the Lean/RfS puzzle.

Strategy Deployment

In Chapter 3, Ian shared an example of strategy deployment with the "Squash Quosh" story. Having an effective strategy deployment process is clearly a critical piece of the puzzle if one wants to make a true Lean transformation. *Getting the Right Things Done* by Pascal Dennis provides an excellent in-depth explanation of strategy deployment. We highly recommend his book to you as a comprehensive explanation of strategy deployment, the terminology used, and how to apply it in an organization. Our intent in this workbook is not to provide you with a detailed methodology for strategy deployment, but to make the connection on the level of importance it plays in a Lean transformation. What is a challenge are the obstacles one faces in doing it well. In other words, you need to do it, but it *is* difficult.

At our first session on strategy deployment, I remember Pascal telling us that it would take us three years to "get" strategy deployment, and he was wrong. It took us more like four years to understand it, and it is still difficult to do well across the whole business. However, at KC, we now recognize just how important a well-implemented strategy deployment process is. We also believe it is impossible to successfully sustain strategy deployment in the firefighting culture that is created by batch logic.

Key Aspects of Strategy Deployment That Helped Achieve a Transformation in the Way KC Operated

First is problem-solving thinking. In the early years, our ability to really understand problems down to the root cause and then develop effective countermeasures to solve these root causes was weak. It took time to build this capability. It connects to strategy deployment because it is about becoming a "learning organization"—one that resolves issues rather than just addressing symptoms, resulting in continually fighting the same fires. Problem-solving methodology will be covered in Chapter 6 under "Rockbusting."

Second, strategy deployment is an application of Plan, Do, Check, Adjust (PDCA) at a strategic level. We knew what PDCA was, but like most companies, we were weak at the check/adjust part. We oversimplified PDCA. It is a simple tool and is easy to understand, but hard to do well. The check/adjust part is about learning, but as well it is about ensuring that activities are directed at achieving the desired results. This is linked to the next aspect of strategy deployment.

Apply a clear focus on the expected outcome of any action. If I do "this," then I expect "that" outcome. Put simply, it is asking the following simple questions:

- What is the outcome you expect from this action?
- Is this outcome in line with your strategic objectives?
- Is the outcome tied to true north and a breakthrough target?
- If yes, how?
- If not, why are we working on this?
- If there is an expected outcome, how can I test it?
- What can I measure in the process to test getting the expected outcome versus waiting to "find out" what happens as an end result?

As part of strategy deployment, it is very helpful if you can find one key in-process measure/focus area as *the* breakthrough. I saw this in action when I had a chance to visit a lift-trucks manufacturing plant in North America. The importance of having a key focus is what I took away from the visit. The focus for that site was on the number of consecutive defect-free lift trucks. They showed me a thermometer chart with the number of consecutive units built without error. The first record on the chart was nintey in 2007. The next year they hit 106, then 195. When I was there in 2010, they were at 366 units. We looked at some of their other business results, and they were excellent as well, but the focus on defect-free lifts was a key enabler for them to deliver on the others. Imagine how this clear and unambiguous focus helped direct everyone's efforts to improve. In the example given by Ian in Chapter 3, this one clear goal was summed up in the code name "Squash Quosh." Everyone knew what the objective was. In the early days of Toyota, it is often quoted that the focus was on reducing the lead time from order to cash.

The Difference between Traditional Strategic Planning versus Strategy Deployment

When people first hear of strategy deployment, they often say, "But we already have a robust strategic process we go through every year." In most companies, it seems to us that what they are currently doing is more an elaborate budgeting exercise; it's often more about setting financial targets the company would like to achieve and budgets for each function. It focuses on the required financial results and tends to be a top-down hierarchical management approach to how the business is run.

Strategy deployment is more about how to do the right things. An example of this was when Ian helped the sales team at a company he was working with to develop their key focus or strategic goal. They went away, as they traditionally did, on a two-day strategic session. They decided their strategic focus would be "to become the best-rated sales team as measured by the Nielsen Retailers Supplier Survey," their reasoning being that if their customers rated them the best sales team, they would be in the best position to develop sales with these retailers. When they presented this back to the senior executive team, they were asked, "But how much are you going to sell next year? That's what you are supposed to come back with after going on a strategy session." Of course, the financials still needed to be done. Now the sales team was focused on what to do to hit their target of being the best-rated sales force—a very different focus than just achieving their sales targets. This focus did, of course, help them not only to achieve their sales targets, but to exceed them.

As I think back on the work we have done at KC with Pascal on strategy deployment, I realize more and more how clever the title of his book is. *Getting the Right Things Done* means focusing on the key breakthrough areas to deliver the outcomes you want to win in the marketplace. This is why strategy deployment is so important. It tells us the outcomes needed in our business and the key process breakthrough areas we will focus on to achieve these outcomes.

Lean Leadership

Lean leadership is a key ingredient needed to achieve a successful Lean transformation. In the early days of Lean/RfS, we had a small core group of leaders plus one senior leader who sponsored Lean/RfS. This meant we could keep moving forward without being bogged down by too many barriers. As time went on, the core team members moved into new roles, the scope of Lean/RfS grew, and the senior leader retired and was replaced. The depth of understanding of Lean and RfS principles and ways of doing things became diluted. It also became apparent that saying the right words of support was not enough. It takes an investment of time for the leadership to build their understanding of Lean principles and then be able to both teach and support their teams on these principles—to live them daily in the way they carry out their own roles.

A whole book could easily be written on this topic—indeed, several already have been! Here is a list of the key elements of Lean we believe helps one become a Lean leader.

- Learn by doing
- Ask questions
- Treat problems as gold
- Learn to see
- Change the KPIs
- Leader standard work
- Go see

For most people, *learning by doing* is the best way for them to learn, retain, and apply knowledge. I've read numerous Lean books, and in almost all cases I've found them insightful. However, they are not the same as leading Rapid Improvement Workshops or personally taking part in problem-solving sessions.

I've seen a great many leaders "support" their teams by providing sponsorship and resources and by offering to remove barriers, but not personally lead by "doing." Leadership behavior changes are a key part of a successful Lean transformation. Becoming a "leader as teacher" is one of the key changes we were taught by Pascal. Personally teaching by leading problem-solving and other sessions with team members is essential. It is not just about showing your commitment; it is also learning for yourself how to successfully use the techniques of problem solving and PDCA. This, in turn, will help others see how to apply them and to understand that this is the way things should be done. Too many senior people tell people to "use the tools of problem solving" but don't actually use them themselves.

Lean leadership also means changing the way you approach daily engagement with people. Team members (at all levels) don't learn when they are told what to do. The command and control structure where leaders tell team members what do stifles thinking and understanding. Making the transition from telling or even guiding (even when very well intended) to *asking questions* and building capability in the team to improve their problem-solving skills is very difficult. Over the course of several years, I have worked hard to make this transition. During this period, we have applied Lean tools to make problems more visible and improve our problem-solving capability.

Here is a small personal example. Visual management is an important Lean tool, so as a senior manager, I could have designed a suitable board and given it to each of my teams. Upon letting each team design their own board, I could easily then have told them, "Next time I come, change the board to look like this," just because I did not like a particular aspect. At first, it was *very difficult* not to give instructions or suggestions for change. But, if I had done so, and although the team would have very willingly made the changes, they may not have asked themselves the reason for the change (a very important question) and understood why I would suggest such a change. I could have sped up the process by telling each team what to change, but their problem-solving and improvement capability would not have developed. The real outcome we wanted to create would have been lost. It wasn't just holding back the instructions/direction that was hard, but it was also what questions I should ask. They didn't come naturally, and it was very frustrating.

As time went on, I found that my learning was helping me determine what questions to ask. This does not mean that leaders do not have a clear role in setting direction. Actually, it is critical, but that is more about where we are going than "Go do this now." Now as I visit my teams on gemba walks, I have gotten more comfortable not giving direction, instead asking a question or two for the team to consider. This will often include a question I write on their visual management board that I'd like the team to answer the next time they take me through what is on their board.

Our old approach tended to hide problems. People worked on symptoms and developing their excuses rather than resolving the root causes. Getting problems exposed is a big change for both team leaders *and* team members. How do you to respond to an exposed problem? Do you say thank you? A famous expression from Toyota is: "No problem is a *big* problem," meaning you will always have problems, but the trick is to expose them and then apply standardized, disciplined, and structured problem-solving analysis leading to root-cause resolution. One needs to learn to *treat problems as gold*, treasures to be welcomed, as they are opportunities to improve.

Another Lean leadership role that needs to happen is what Kimberly-Clark terms "learning to see"—seeing what others can't see and identifying the opportunities for further improvements. One can then set direction through strategy deployment to address those opportunities that are potential breakthroughs. Teaching the team *learning to see* for themselves is an important Lean leadership skill.

Here is an example from KC. After we had applied Lean/RfS to many of our plant production processes, we selected a distribution center to apply Lean/RfS principles. We held a 5-day Rapid Improvement Workshop. Applying the Glenday Sieve to demand through the distribution center, we discovered, as usual, that only a small number of customers represented 5% of the volume. One customer in particular took over 25% of the volume on their own. So we focused on developing a green-stream flow for that customer. The team did an excellent job putting together a specific zone for the SKUs the customer ordered. They developed a routine fixed time flow for products produced at the plant and others that were imported from other locations to match fixed deliveries to the customer. It worked brilliantly!

I had the opportunity to visit the distribution center (DC) four years later. It was the first time I had been back to the site since the original workshop. I was very happy to see that the green-stream zone and process for this customer was still in place. But I also noticed the process was pretty much the same as what was implemented during the workshop four years earlier. I wondered why that was. As we stood on the floor by the green customer zone, I observed the operation. I asked the DC leader what form of waste he could see. How much inventory was in the green zone? He could see that there was an opportunity to further reduce inventory, but what I wanted him to notice was the motion waste. There were lift trucks moving without product, and there were some travel distances that appeared far too long to me. Four years ago, reducing the distances traveled by lift trucks was seen as a major improvement to what had been before. But now I could imagine a much smaller green-stream zone with less inventory and shorter distances traveled. I challenged the team to think about how they could improve the process further. They had a far better operation after the application of Lean/RfS. It was good that they had kept the disciplines and not fallen back into "the old ways." However, it was also a disappointment to see little further improvements since the first workshop.

This story illustrates a couple of points. How do you "learn to see"— to see flow gaps, waste, and other opportunities to continually improve? The second point is being convinced that there is always another future state. More opportunities always exist; we just need to be able to see them. It takes time and experience to see waste and flow gaps in a process. It is about critically observing a process—looking where the flow is interrupted and where effort or resources are being wasted. Management at Toyota draws a circle on the floor for people to stand in to do this observation. It applies to any process, not just manufacturing, but wherever the work is done. Try it in offices and laboratories. I'm still amazed at the opportunities I didn't see before and how obvious it is once I do see them. It takes time to do these observations, and time is something leaders often think they don't have enough of.

That's why we found that *leader standard work* was critical in helping achieve Lean leadership. It helps leaders manage what they should be focusing on, to more effectively use the time available to them, and to understand how to prioritize the most important tasks you should be doing as part of your job. Then establish a fixed routine—ideally weekly or even daily—for these tasks to ensure that they get done, with agreed upon rules so that this time is not disturbed. The rest of the time is available for less important aspects of the job. You could call these "green" and "red" activities, which is what we do.

Earlier in this section, Ian described an exercise developed at Colman's of Norwich to help people identify their green and red tasks. At KC, we agree that green tasks for leaders are generally those associated with the motivational, encouraging, and development aspects of the job that people know they should be doing yet somehow never quite managed to achieve on a regular basis. We also realized the importance of making process checking part of the green tasks. What are the key processes? Who owns them? Ask process owners if their process is healthy. Personally observe the process to identify improvement opportunities. Then, teach the process owner "learning to see." It is all part of leader standard work.

Many senior managers find leader standard work hard to accept and do at first, as it is not how they normally manage their time. Unfortunately, for many, the usual way involves letting the less important aspects of the job get in the way and far too much firefighting. Breaking out of this way of working is easier with a fixed repetitive routine to follow. And once the pattern is established, people find they like it, including the team members the leader is responsible for.

Toward the end of our first workshop, I remember asking Ian, "As we implement Lean/RfS, what measures, or KPIs, should we put in place?" "I'm glad you brought this up," he replied. That set us on a path that is still underway. What are the *right* measures? What KPIs *need to change?* Since the logic of Lean/RfS is different from traditional batch logic, at least some KPIs need to change. The "old" measures support the old logic. There is a saying that one definition of insanity is doing the same thing and expecting a different outcome. If you keep the same KPIs and expect different outcomes, you are insane!

One key measure to put in place—no matter what the process—is conformance to plan. Hit the plan on time for the "green" activities. The discipline around this KPI is critical, or the ability to work in a stable flow environment will fall apart. This change was significant for KC. Previously, output units × time was calculated into an efficiency percentage. This was the number one measure in manufacturing. In addition to making conformance to plan a key metric, production efficiency needs to be less emphasized. Consistently hit the plan and we guarantee production efficiency will improve. At least, it always has at KC!

Total system performance measures need to be emphasized and valued over component or functional performance. Again, this is true for any process. A good example would be to focus on total delivered cost versus unit production or transport costs.

Another question to ask is how well your measures connect to strategy deployment and key breakthrough outcomes. Are they directly tied to them? Do they support them? The stronger the connection here, the clearer and more aligned the teams will be as they do their work. We have found that having the debate on what the right measures are is as important a part of the process as deciding what are the right KPIs. It helps people understand the process itself better. We have found that KPIs usually evolve as people further develop their understanding of Lean and how it applies to process improvement. Therefore, it is difficult to say, "These are the right measures."

The last element of Lean leadership to explain is what we at KC call *go see.* It combines many of the other elements and is something that Jim Womack and Dan Jones often talk about doing. Get a group of people together to walk the value chain, or at least a part of it. It could not be simpler, yet it just is not done enough.

Here is an example of a recent highly successful "go see." KC was working with a customer to determine some shared improvements between our companies. Great progress had been made over the past several years to improve the flow of product from KC through the customer. As part of this, we agreed to jointly walk the process from the customer store shelf back to the manufacturing process. We saw many things we would not see by meeting in a conference room. The number of touches the product went through to get from manufacturing to the consumer was much higher than we imagined. We were able to see and then ask, "Why do we do it this way?" in many areas. We questioned the whole design of the product configuration on the pallet and designed a different layout that used less packaging. It also improved freight efficiency, resulting in better flow and velocity through the supply chain. The other important aspect of "go see" is the connection to the people doing the work. It builds a connection between leadership and employees in a way that cannot be replicated. What better way to reinforce values, see barriers that need to be removed, ask questions, and say "Thank you!"

Four Rules of Lean

Something that has helped KC in their Lean transformation is an understanding of the "four rules of Lean." This is based on the 1999 *Harvard Business Review* article by Spears and Bowen, "Decoding the DNA of the Toyota Production System," plus further teaching on these rules from Pascal Dennis. They are totally aligned with green-stream and red-stream flow principles. Here are the four rules:

1. All work should be highly specified in its content, sequence, timing, and outcome.
2. Every customer/supplier relationship should be direct, binary, and self-diagnostic.
3. The pathway for each product and service should be simple, prespecified, and self-diagnostic.
4. Problems should be solved using the scientific method at the lowest level supported by a capable teacher.

The first three rules are "design" rules. They are rules to follow when designing your processes. The last rule is about how to tackle issues within your processes when they arise. Here are some examples of how these rules and Lean/RfS tie together to help develop Lean-thinking solutions.

In a green-stream flow, there is a connection between different production assets as in a customer/supplier relationship. For instance, in our Baby Wipes plant, when we consume parent rolls, there is a direct connection to the base machine (the supplier) through a supermarket. The base machine produces parent rolls as directed by the trigger signal of consumption in converting. Only converting is scheduled. Previously, with batch logic, both assets were scheduled with variable levels of work-in-progress inventory between the assets.

Customer order sourcing is another example of where we have learned to understand the relationship of the green-stream flow and the four rules of Lean. Previously, KC had set up a dynamic order-sourcing IT program to "optimize" costs. When an order came in, the computer would review it to see what shipping distribution center (DC) would be the lowest-cost option. We "saved" millions of dollars by moving orders around to a location that could deliver it to the customer at the lowest calculated transport cost. Our old thinking and logic led us to believe we were saving large sums of money every day, when actually, it was adding complexity and chaos to the supply network. For instance, we have regional DCs in Chicago and Kansas City. Chicago is 300 miles from St. Louis and Kansas City is 250 miles from St. Louis. From a customer-sourcing perspective, both locations have reasonably similar shipping costs to St. Louis. Order-sourcing logic would review the order and select the optimal location based on the configuration of products on the shipment. By doing this, we masked the customer-demand signal as well as continually switched demand between the two DCs. That resulted in variable workloads on both DCs, meaning extra but hidden costs. With both Lean/RfS and the rules of Lean, the answer is one consistent shipment point. When there are problems, which there will be at times, understand the root cause and fix it. Don't just switch delivery points and hide the problem.

This does not mean that distribution optimization models should not be used. However, they should be tied to a more strategic review of the network to design a repetitive, routine pattern. Think of it as a green stream for the whole network flow process. Demand shifts will occur, and at some frequency these need to be remodeled to create the best flows for the current situation. We would recommend doing this on a quarterly or twice annual basis.

The four rules are a guide for how all your processes should be designed and working. The hard part will be going from your current condition to a future state where you are operating under the four rules of Lean.

Summary of Chapter 5

Like a real jigsaw puzzle, the straight edge pieces help you frame up the puzzle, giving it structure and shape to help complete the whole picture. There are Lean/RfS straight edges. You will be creating material green-stream flows, managing non-value-adding blues and reds, changing KPIs and costing models, plus improving all processes by applying Lean/RfS principles. However, on their own, these are not sufficient. Strategy deployment as well as developing a deep understanding and commitment through Lean leadership learning are key straight edge pieces you will need if you want to achieve a successful Lean transformation. It is a long and at times difficult journey. It is also highly rewarding, exciting, and motivating, and it will help you achieve levels of improvement previously thought impossible, not just in performance but also in behavior of people and how they work together. Are you ready for the challenge?

In Chapter 6 we cover the specific Lean/RfS tools and techniques that help fill in the center of the puzzle. There are other Lean tools—such as SMED, TPM, and kanbans—that you will also need to use. We will not describe them here, as they are adequately explained in any Lean book on the subject.

Chapter 6

The Lean/RfS Center Pieces

Most people are familiar with Lean tools like SMED, 5S, VSM, TPM, and others. (See Glossary at end of book.) It is these tools that many companies focus on in their Lean initiatives. These equate to the center pieces of the Lean/RfS jigsaw puzzle. They are important if we are to understand the whole picture, but they are not a critical part of achieving the logic change of batch to flow. Indeed, many organizations successfully use these techniques while still using batch logic in their planning systems. However, it does mean that they have the issue of a different plan each week, which then usually changes, and this makes sustainable improvements difficult to achieve. But that does not mean that these tools shouldn't be used. They are all good techniques aimed at helping improve performance, which for many people means increasing line efficiency. This is where Lean/RfS differs from what many people would consider the right thing to focus on: line efficiencies. With Lean/RfS there are two key points of focus:

- Conformance to plan, not efficiencies
- People, not machines

The Lean/RfS center pieces are specifically aimed at ensuring that the plan is met. This is paramount. Second to this is identifying and then resolving the issues that frustrate people, especially operators. These are the issues that make their working lives harder than they need to be. Of course, the overall objective is to increase performance, and that means line efficiencies. However, that should not be the prime focus for people looking to improve performance, as efficiency is an *output* KPI affected positively by Lean/RfS.

Why Is Conformance to Plan the Prime Focus?

Firstly, whenever data has been collected on where the most requests for plan-changing come from, it is *always* from production itself. If something goes wrong with not enough being made, or if all goes well and too much is made, it's back to the planner to sort it out, and the plan is changed. It is a constant source of variability. Production should be focused on making the plan—not too little, not too much.

Secondly, precisely because it's easy to change the plan to compensate when things don't go to plan, there is little or no root-cause identification and resolution of what went wrong to attempt to stop it from happening again. There is a lot of symptom-solving going on—otherwise known as *firefighting*.

The third reason is around inventory and customer service. If one wants to get inventory levels down, to have JIT deliveries, and still have high customer service, one needs frequent reliable manufacturing. Make the plan—every time. The production schedule should not be seen as a guideline on what gets made; it should be *exactly* what is required.

Consider the railway systems. In the UK, the train service is seen as unreliable and expensive, with frequently late or cancelled trains. The high cost of fares is a direct factor caused by relatively low customer numbers, precisely because it has poor customer service. Contrast this with rail networks in Switzerland or Japan. They have extremely high punctuality rates, fares are considered a good value, and trains are clean and enjoy high passenger numbers. And, whenever a train is late, operatives carry out investigations into the reasons why. So many companies seem to be more like the UK rail network when they should be striving to emulate those in Switzerland and Japan—and it can be done.

Rockbusting

The term *rockbusting* originated at Kimberly-Clark. As part of implementing a green-stream fixed cycle, a team was looking at ways to ensure the green-stream plan was achieved—the focus being *make the plan*. Someone made reference to an old Lean analogy of the river of rocks, with inventory representing the water covering the rocks. The original analogy suggested dropping the level of inventory to expose the rocks and then tackle those problems first. Following this, the inventory could be dropped again to expose the next rock, and so on until all the rocks were broken. The team at KC saw rocks as disrupters of the green stream that were stopping it from flowing smoothly. Rockbusting consisted of actions to get rid of these disruptive factors, with the logo and slogan borrowed from the film *Ghostbusters* (see Figure 6.1). The term just stuck.

The second focus of Lean/RfS is people

Production is so often obsessed with line efficiencies and breakdowns. The biggest losses of time are seen as the most important issues to address. The longer the breakdown, the more important it is. With Lean/RfS, *people* are seen as more important. It is not the biggest stoppages that are important, but those that are small and most frequent. These small stops cause operators stress and frustration, and make them work harder than they need to. When visiting factories around the world, we are often amazed at how committed and hardworking the operators are, trying their best to produce as much as they can. They work hard to overcome the little stops, many of which don't even get measured in any analysis of line issues. Or if they do, they receive little attention from technical and management staff to help fix them, because they don't add up to the biggest loss of time overall. The operators are left to cope—it becomes part of their job.

Yet these small stops can often be relatively easy to fix. Remove the stops, and operators have more time—more time to take care of their machines, listen to them, recognize when something doesn't sound right, or recognize a small issue developing when, for instance, a new pallet of packaging material is introduced.

In our experience, when one focuses on resolving the frequent small stops, a curious thing happens. The bigger breakdowns are also significantly reduced, and not just because operators are less frustrated and more in tune with their machines. Short stops are bad for machines as well. Frequent stop/starts increase loads on mechanisms, such as chains, clutches, and gearboxes. Reducing the short stops is likely to also reduce wear and tear on these types of machine components, meaning breakdowns are less likely to occur.

The Lean/RfS center pieces are focused on these aspects of improving overall performance.

Line 31 Rockbusters

Figure 6.1   Rockbusting logo.

Rockbusting is aimed at ensuring high conformance to plan and consists of two separate activities. The first is monitoring hourly outputs versus plan. Was each hour good, i.e., at or above planned output, or bad, i.e., below planned output? The second activity is to have a structured problem-solving process to resolve what is causing the bad hours, that is, the rocks. It's not about the biggest breakdowns, which usually already have a process in production to identify and address these. Rockbusting is about looking at the most frequently occurring issues—the short stops. Each short stop may only be less than a minute, but if they occur frequently, the time adds up, resulting in a bad hour. One often finds that good hours balance bad hours, so across a shift outputs can look OK overall, but hour by hour there can be quite a difference between the hourly output rates. This monitoring should be a visual chart placed next to the line and filled in by operators using simple green and red indicators. It does not need to be numbers, as it is *not* measuring line efficiency. It's a quick way to identify when the line is running well—so everything is OK—or not so well with red hours. The aim is to then identify the reasons why the line is experiencing the issues. Two examples of actual rockbusting monitoring charts are shown in Figure 6.2.

This is a simple example, easy for people to fill in and, more importantly, easy for team leaders, engineers, and production managers to see quickly when the bad hours are occurring. The intention is that they can then ask the operators what issues they're experiencing while it's still fresh in their minds. The operators can show what the issues are. It is not unusual to find that an issue repeats quite frequently for a time and then goes away, so operators will have a better chance of identifying the issue while it's happening.

An example that we have seen many times is with any cardboard component. A particular pallet of cardboard boxes will, for no apparent reason, run badly with repeat jams and smashups. The run will then return to normal when another pallet of material is introduced. Invariably, the cardboard supplier gets the blame for providing poor quality packaging materials. So why is it possible for one pallet of cardboard materials to run worse than normal? The answer usually lies in how the cardboard boxes have been handled and stored. Reasons we have seen include:

- Poor stock rotation in the material warehouse, meaning that every so often an older pallet of boxes is found and sent to the line. Old cardboard does not run well.
- Restricted warehouse space, resulting in a few pallets of boxes being held in a temporary area situated below the warehouse's main heating unit. Boxes exposed to heat do not run well.
- Backup of trailers being unloaded into the material warehouse. Sometimes trailers were left to be unloaded, sometimes days later, in the southern United States, where summers are characterized by high temperatures and humidity. Boxes exposed to these extremes definitely do not run well.

Productiecontrole

Uur	1	2	3	4	5	6	7	8
DINSDAG	06:00–07:00	07:00–08:00	08:00–09:00	09:00–10:00	10:00–11:00	11:00–12:00	12:00–13:00	13:00–14:00
Doel	4320	4320	4320	4320	4320	4320	4320	4320
Aantal	4320	4400	4320	3800	5000	4320	4460	1050

Uur	1	2	3	4	5	6	7	8
WOENSDAG	06:00–07:00	07:00–08:00	08:00–09:00	09:00–10:00	10:00–11:00	11:00–12:00	12:00–13:00	13:00–14:00
Doel	4320	2880	4320	4320	4320	4320	4320	4320
Aantal	4320	4400	5000	4550	3800	4460	1810	1050

Uur	1	2	3	4	5	6	7	8
DONDERDAG	06:00–07:00	07:00–08:00	08:00–09:00	09:00–10:00	10:00–11:00	11:00–12:00	12:00–13:00	13:00–14:00
Doel	4320	4320	2880	4320	4320	2880	2880	4488
Aantal	4112	593	3230	5280	4110	2180	2475	950

☺ Beter dan doel   ☹ Erger dan doel

Figure 6.2 Example of a rockbusting board.

The rockbuster board helps improve performance in two ways. Firstly, it is an example of applying the Hawthorne effect, as described in Chapter 5 of this workbook. It provides highly visual hour-by-hour monitoring of performance next to the line that the operators themselves complete. This tends to increase performance simply by being there. Secondly, it works as a "trigger" for people who can help resolve rocks. It shows when it is a good time to talk with the operator—during red hours—so that they get firsthand feedback on what the issue is. In many factories, operators are often asked to fill in logs—either in books or computerized versions—listing what issues they have. In our experience, these do not aid root-cause resolution of issues, as invariably when one reads these logs the same issues keep reappearing at different times. The logs are completed, yet there seems to be little structured analysis to identify and resolve the root causes of the frequent short-stop issues. Getting firsthand verbal feedback from operators plus a "look see" at the line while the short stops are occurring is much more likely to help resolve repeat issues.

Figure 6.3 is another example of a rockbusting monitor that is more detailed than the first. The same green and red is used to highlight good and bad hours filled in by operators. This factory also wanted to track cumulative outputs and had space for any comments the operators wanted to record.

At first, Ian did not like this example, as it seemed to be mixing too many items on one sheet. However, it was well received by the operators, as it had done away with several other sheets they previously had to complete. They did not *have* to add comments—only if they wanted to share their views on what the issues were. Two additional benefits came from this format. The team leader told me that all the operators would look at it repeatedly during the shift, to check on the cumulative figure, to check if they were on track to make the required shift output, even if they had had some issues earlier in the shift. And they invariably made the planned output for the shift. It was the Hawthorne effect.

The team leader also started to notice how often the first hours of any shift were red. The pattern can be seen in this example. Everyone knew the first hour or two of a shift was a settling-down period. This had not been recognized before, as previously output was only measured over the entire shift, not hour by hour. Operators, engineers, and team leaders discussed why this was, and the consensus was that during the first hour of their shift, operators generally adjusted the machine settings to what they considered to be the "correct" ones, then taking the next couple of hours to getting them "just so."

Of course, every shift had its own view of what were the "right settings." There were no standard settings across all shifts, and previous attempts to achieve this had failed, mainly because each shift *knew* deep down that they had the right settings, so they were not going to change them, no matter what they were told. But now they had a graphic illustration of how wrong this was. Getting them to agree on cross-shift settings became much easier, with the goal of achieving green for the first few hours, when previously they were typically red. This proved a surprisingly good motivator; people like a goal they can understand, and it provided some healthy competition between shifts on who could achieve it first.

It doesn't matter what the hourly monitoring looks like; we have lots of different examples from different companies. What is important is that it is visual, next to the line, easy to fill in by operators, and is looked at several times across a shift by team leaders and *every* manager who happens to walk past. This makes people realize how important it is.

The second aspect of rockbusting is to have a common problem-solving process in place to identify and resolve the root causes of rocks. Nearly all companies have done some training in problem-solving tools and techniques, so people know what they are. The problem is that they don't use them. It's much more fun to be a firefighter and solve the symptoms than go through the exercise of PDCA or the Five Whys.

In bringing together several large global FMCG companies to discuss their experiences in implementing Lean/RfS, the biggest common issue was a lack of problem-solving expertise in their shop floor leaders. They'd had the training

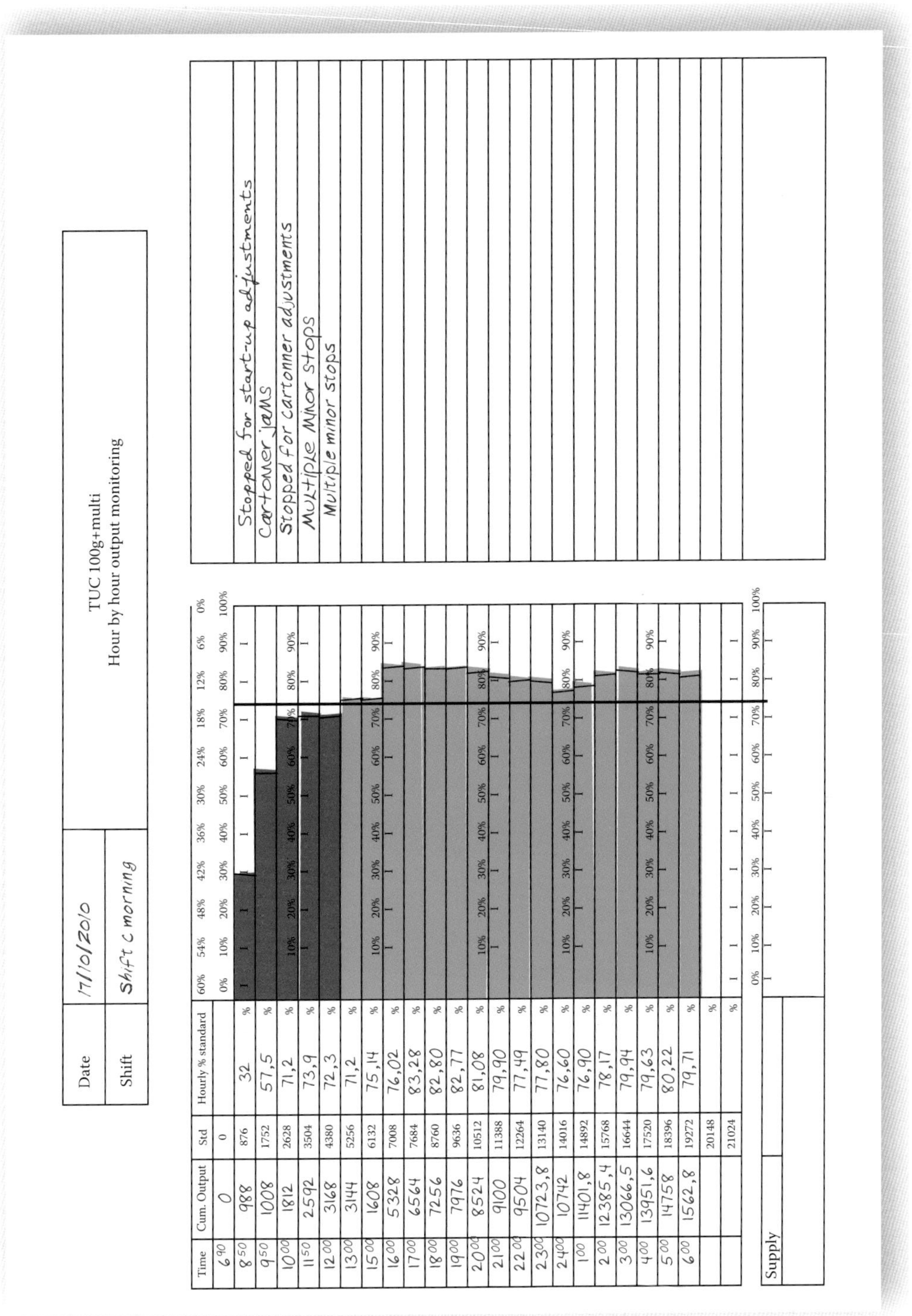

Date	17/10/2010
Shift	Shift C morning

TUC 100g+multi
Hour by hour output monitoring

Time	Cum. Output	Std	Hourly % standard
6:00	0	0	
8:50	988	876	32
9:50	1008	1752	57,5
10:00	1812	2628	71,2
11:50	2592	3504	73,9
12:00	3168	4380	72,3
13:00	3144	5256	71,2
15:00	1608	6132	75,14
16:00	5328	7008	76,02
17:00	6564	7684	83,28
18:00	7256	8760	82,80
19:00	7976	9636	82,77
20:00	8524	10512	81,08
21:00	9100	11388	79,90
22:00	9504	12264	77,49
23:00	10723,8	13140	77,80
24:00	10742	14016	76,60
1:00	11401,8	14892	76,90
2:00	12385,4	15768	78,17
3:00	13066,5	16644	79,94
4:00	13951,6	17520	79,63
5:00	14758	18396	80,22
6:00	1562,8	19272	79,71
		20148	
		21024	

Supply

Stopped for start-up adjustments
Cartonner jams
Stopped for cartonner adjustments
Multiple minor stops
Multiple minor stops

Figure 6.3   Another example of a rockbusting board.

1000 LINE PROBLEM SOLVING TOOL - THE 5 WHYS?????

Date............... Shift............... Time............... Operator...........................

Step 1 - State the problem: Continuous stream of jars feeding out of the capper without caps. Impacts time and money

Step 2 - Ask why the problem identified happened. List ALL of the answers you can think of in the left hand column, leave plenty of space between the different answers. Then ask why again and list the answers in the next column.

Ask why the problem above occurred?	Ask why the problem on the left occurred?	Ask why the problem on the left occurred?	Ask why the problem on the left occurred?	Ask why the problem on the left occurred?
1) Distorted cap jammed in chute (after no cap sensor)	distorted in moving (supplier distorted) damaged in process	wear and tear jammed in chute sensor	operator error operator error overfill hopper	Lack of training/knowledge
2) No caps in chute	dirty sensor (no cap) dirty reflector faulty sensor	dust from wad/coffee	no crate to use folded down creator error	
3) Spring in chute opener failed	spring snapped spring stretched broken bracket	quality of spring wear and tear wear and tear	supplier quality issue lack of maintenance chuck fault	other concerns/not done/higher priorities lack of training
4) Malfunction with the escapement	piston fault bracket fault incorrectly fitted	failed operator error tech error	wear and tear lack of training/knowledge	lack of maintenance
5) Trapped wad/diaphragm in chute	wads come loose	distorted cap excess air along jet stream conveyor	set up incorrectly after last run supplier/ai error	system error

Figure 6.4 Example of a problem-solving process.

and encouragement, but it still didn't seem to gel for people on a day-to-day basis. Rockbusting is intended to install a standardized problem-solving methodology that is visible on the shop floor and can be followed, with the team leaders facilitating the process. It almost doesn't matter what the standardized problem-solving approach is as long as everyone understands and uses it. It requires senior management to support it and ensure that it happens. They should personally take part in problem-solving groups to show commitment, interest, and active involvement in helping solve rocks. They should not take over from team leaders, as it is the team leaders who need to be the ongoing facilitators of the rockbusting process.

The example shown in Figure 6.4 is quite simple. In some companies, an ongoing audit is also implemented to ensure that the process continues. This example is using the Five Whys. It provides a visual framework, a means to get everyone's input and to force people to think a little deeper about why something is happening, to go beyond their first knee jerk reaction. The point of this is to ensure that team leaders follow a process with a group of people, including operators, to attempt to solve rocks. We suggested that this process works on only one rock at a time. Focus on one, fix it or park it as "unbreakable," and then come back to the next rock. Don't have a long list where nothing seems to get resolved, with the only action being the amendments of the completion dates.

Rockbusting can be a difficult process to sustain, precisely because it is disciplined and structured, and that's not the way most frontline leaders usually operate. They are often the most reactive of firefighters—and they're good ones as well. They need to become proactive problem-solving rockbusters to resolve the recurring issues on the line that frustrate the operators, including having to work harder than is necessary. A monitor that can help get frontline leaders engaged is *mean time between stops*. Measured across a shift, this is time available to run divided by the number of stops. It does require some form of automated line monitoring in place to be able to capture and calculate the figure; however, many companies have monitoring in place that can be adapted to provide that.

It can be very revealing when first calculated. When the figure is first calculated, it is often around six or seven minutes, which can surprise people, as it's so low. They might not have realized that the short stops were quite so frequent. High-speed lines that have one of the machines on the line stop every six or seven minutes are not performing well. Frequent stoppages are not good for the machines or for the operators, who have to deal with whatever caused the stoppage. Having team leaders focus on increasing this number is a good way to get them engaged and eases them away from just thinking about the line efficiency measure. They can then start to identify and eliminated the reasons for these most frequent stoppages, in other words, bust the rocks. Any improvement in mean time between stops is seen by operators as positive. And if the number starts to rise, then increased line efficiency is bound to follow.

A word of warning: We often find that automated lines have been designed with "buffers," such as accumulation conveyors between the various pieces of equipment on the line. These cover up—deliberately—the short stops, so that check sheets may be required on the line to enable the operator to put a "cross" every time a short stop occurs. In this way, it's possible to get a reasonable analysis of just how many short stops are happening in order to calculate mean time between stops.

Does Rockbusting Deliver Results? You Bet.

The first example of rockbuster results comes from a branded confectionary company (see Figure 6.5). The results really surprised people—they did not expect to see quite such a big jump in outputs. A combination of Hawthorne effect with actual rock fixing, together with the stability of fixed green-stream cycles that lead to economies of repetition, really can have an impact that is as immediate and significant as the example shown here. Active support and involvement from senior management to help make it work was also a contributory factor here. Figure 6.6 provides some results and comments from Kimberly-Clark.

Schedule Breaks

A schedule break is where the agreed plan is changed, for whatever reason. Planning needs to implement a schedule-break process to track the reasons for

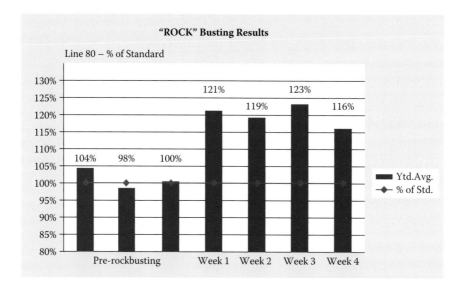

Figure 6.5 Rockbusting results chart.

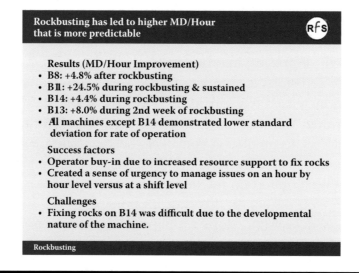

Figure 6.6 Some KC rockbusting results.

any changes, followed by problem-solving techniques to identify and resolve the root causes—the planner's version of rockbusting. It is inevitable that schedule breaks will occur at times. What is required, however, is a means to understand why and learn from it so that it happens less and less.

What is the agreed plan? The green stream is fixed for the length of the cycle, so that's a given. But what about the red SKUs? Generally, people agree on fixing the plan around midweek in the week before production for the purposes of monitoring schedule breaks. Yes, the reds are more volatile and therefore likely to change, but that doesn't mean lessons on *why* reds change can't be learned. It *should* be possible to have a stable plan for one week. For some industries—especially those working with short shelf life—it may be more difficult than for the industries able to hold some finished goods inventory to cover for sales variations of the red codes.

Therefore, what is required is a monitoring mechanism to plot schedule breaks, together with the reasons why they happened. This should be displayed on a board in the planning department. As with rockbusting, the specific format and detail should be agreed upon by the people who will use it. When schedule-break tracking was first started at Kimberly-Clark, a problem arose. The chart showing the number of schedule breaks per manufacturing site was produced, with good intentions, as part of the monthly KPI pack that was circulated to all sites and senior management. It was seen as a KPI, with sites having more schedule breaks performing worse than those with fewer—*wrong!*

Monitoring schedule breaks is part of problem solving and becoming a learning organization. It is monitored to trigger an analysis of why the break happened and what can be done to stop it from happening again. Having more or fewer schedule breaks is not the issue. What *is* important is whether a site is investigating the reasons and taking action to prevent them from happening again. KC instigated a program to educate people on the reasons why schedule breaks needed to be monitored and data collected as proof of the reasons why. It ran under the slogan:

Schedule breaks are not BAD—not knowing WHY is bad.

How KC categorized the areas responsible for initiating schedule breaks is shown in the pie chart in Figure 6.7, which illustrates the typical percentage breakdown.

KC used the term *deeper dive* to describe examining in more detail the reasons for breaks within each main group. They produced a Pareto chart to guide them on where to focus their efforts in understanding and resolving the root causes. A typical Pareto for suppliers is shown in Figure 6.8.

The reasons causing schedule breaks need to be exposed and understood. It is usual for people to say, at first, that it's beyond their control or influence.

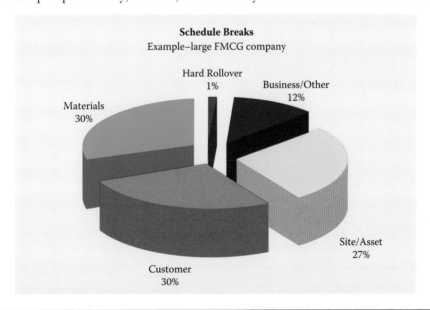

Figure 6.7   Schedule breaks: typical percent breakdown by category.

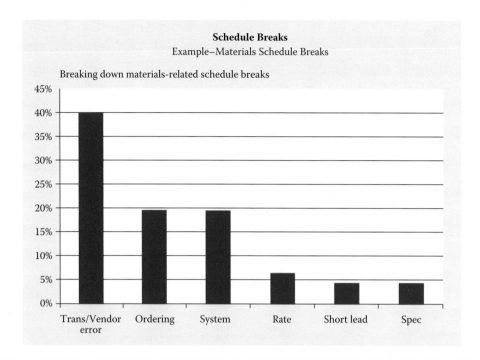

Figure 6.8 Schedule breaks: Pareto chart for material issues.

That could be true of some of the reasons, but not all. So, start with the things you can resolve. The fixed repeating green stream with a buffer tank to protect this schedule *should* make it easier to resolve issues that disrupt the plan. There will be, at times, genuine reasons why the green stream needs to change—it will happen. However, many of the reasons why plan changes currently occur can be controlled so that in the future they will no longer disrupt the plan.

How to Measure Conformance to Plan

The way that conformance to plan is measured with Lean/RfS is called TIP/TOP. It is a play on the English expression to be "tip top," meaning that everything is excellent.

TOP is Total Output Performance. It is calculated as follows:

$$\frac{\text{Sum of total volume of production made}}{\text{Sum of total volume of production planned}} \times 100\%$$

Table 6.1 shows an example of how to calculate TOP.

It is possible to get over 100% with TOP. It measures if the plan is being met in overall terms and ensures that the total volume in the cycle is within the demonstrated capacity. Meeting the plan needs to be at an item level as well as overall volume level. Overproduction at the item level should be seen as undesirable as underproduction.

Table 6.1

$$\frac{\text{Sum of total volume of production made}}{\text{Sum of total volume of production planned}} \times 100\%$$

For example:

Product	Planned quantity	Actual quantity made
A	100	90
B	150	160
C	50	40

$$\frac{(90 + 160 + 40)}{(100 + 150 + 50)} = \frac{290}{300} \times 100 = 97\%$$

Table 6.2

$$\frac{\text{Sum of total volume planned–sum of differences actual vs. planned by item}}{\text{Sum of total volume of production planned}} \times 100\%$$

For example:

Product	Planned quantity	Actual quantity made	Difference Plan versus Actual
A	100	90	10
B	150	160	10
C	50	40	10

$$\frac{(100 + 150 + 50) - (10 + 10 + 10)}{(100 + 150 + 50)} = \frac{270}{300} \times 100 = 90\%$$

TIP is Total Item Performance. It is calculated as follows:

$$\frac{\text{Sum of total volume planned} - \text{sum of differences actual vs. planned by item}}{\text{Sum of total volume of production planned}} \times 100\%$$

Table 6.2 shows an example of how to calculate TIP. Note that all differences are taken as absolutes, not plus and minus the planned quantity. Underproduction and overproduction both count as a difference to the plan.

TIP measures how well the quantity for each item in the green stream is being met. It is a much harder measure for production to achieve a high result. Initially, it is not unusual to achieve less than 70% for TIP. One should expect to see improvements in TIP as the fixed cycles become more stable. Consistently achieving 95% plus TIP is the target.

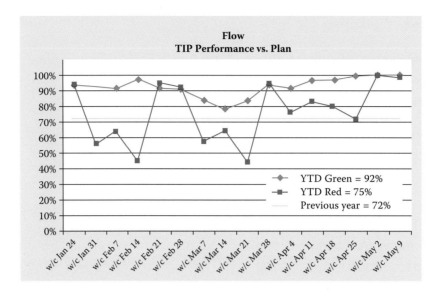

Figure 6.9   TIP performance vs. plan.

Many people believe it would be impossible to hit 100% TIP. That is, producing *exactly* the quantity required for all SKUs in the weekly schedule, with red as well as green items. The graph of TIP performance in Figure 6.9 is from a company that implemented weekly cycles for green and red streams. They were able to run each stream on its own dedicated equipment. The green stream was fixed sequence/fixed volumes. The red stream was variable in terms of products, sequence, and volumes. However, the company worked to keep both streams stable. No changes were made to either plan once issued. Both streams were monitored using TIP/TOP.

The green stream achieved an immediate improvement in conformance to plan over the previous year's average. The red stream showed a great deal of variability in TIP. This was to be expected.

There was a dip in performance in the green stream after a few weeks. This was mainly due to overproduction of items—the result of the economies of repetition. Having achieved higher efficiencies in the green stream, the production people were at first reluctant to stop production when they met the plan. They produced more, making the TIP result go down. As they controlled overproduction, conformance to plan crept up, reaching an amazing 100%. The red stream continued to be erratic, as expected. However, it suddenly also shot up to 100%.

The reason was that the green-stream people had, of their own accord, gone to help the red stream. When asked why, they said it was so that all of the factory could hit 100% TIP/TOP. This would never have happened before Lean/RfS, and it's a great example of how behavior can also be changed by changing the logic from batch to flow.

Achieving 100% conformance to plan at an item level in both the green and red streams was beyond what anyone at this company thought possible. Yet they managed it just fourteen weeks after starting their green and red streams.

Time versus Quantity

When first implementing green and red streams, it is usually better to start running green stream to time and red stream to quantity. It is something of a shock to people, as production had always been planned and run to quantity. People are quick to point out that running to time is likely to mean a poor TIP result—and they are right. So why advise to run to time for greens?

Running to time means that one can fix the time when changeovers, cleanups, and other similar activities happen, so that they occur at the best time, when the right resources are available. It also means that technical experts can ensure that they are there to help reduce the time taken and reduce ramp-up losses through the application of SMED improvement techniques. It has both positive and negative implications for material supplies. Fixing the time means that ensuring materials are available at the right time becomes easier. The downside is that production may make more in the same time, meaning that extra materials need to be available. People can relate to fixed time: Knowing that things happen every week at the same time is easy to remember. This can mean that economies of repetition happen quicker, and creating the phenomenon of economies of repetition is one of the prime objectives of fixed cycles.

There is also a psychological reason. It will be a different process than what was previously applied, so people can appreciate that Lean/R/S really is a change to what they do. Lastly, once they've got their minds around the idea, it will make more sense to people, because if the line runs better—through economies of repetition—then more product is made, and people *like* that. It makes the operators feel good. Management like it too, as they see their efficiency and fixed-cost figures improve. Of course, everyone is worried about what happens if the line runs poorly. You get less production. But won't that potentially cause a customer service issue? That is one of the reasons why there is a buffer tank—to absorb variability in outputs.

There are also contingency plans within the rules that describe what to do if the inventory falls below the lower buffer tank limit. However, the evidence shows that it is far more common to have overproduction when running to time because of economies of repetition, improved changeovers, better material availability, and the impact of rockbusting. It is common to see TOP go above 100% and TIP to fall as a result when running to time, so much so that at times there should be a schedule break in the green stream as finished-goods inventories rise outside the buffer tank and beyond what is acceptable. Overproduction is as bad as underproduction; in fact, many Lean experts consider overproduction to be a bigger waste than underproduction—and we would agree. Running to time is a first-stage process, in order to get fixed cycles running and economies of repetition occurring so that improvements in outputs and efficiencies can happen.

The alternative would be to run to a fixed quantity. When first implementing green streams, this has a number of disadvantages, the key disadvantage being that operatives still want occurrences like changeovers and delivery of materials

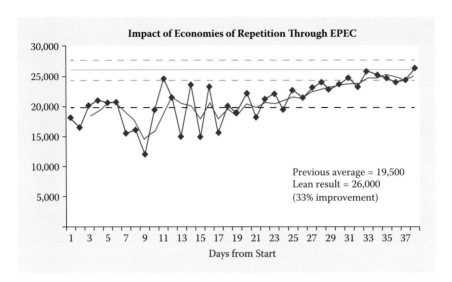

Figure 6.10 Impact of economies of repetition through EPEC.

to happen at fixed times to create stability, reliability, and routines. If the lines run better, then to achieve this means stopping production when the fixed quantity is made and then waiting until the next SKU is due to be run. Pulling production forward means that materials may not be available, and the changeovers start to happen anytime—exactly as happens in a batch logic situation. Nothing has changed. If production runs poorly, then the line should be changed over at the scheduled time for the next SKU; otherwise, the routines will be ruined again. This means that production will only get the downside when things run poorly and not the upside when they run well. This scenario has major credibility issues with nearly everyone—but especially operators—when it is explained to them. Stopping the line and waiting just does not make sense to them.

However, with Lean you *do* want to run to exactly the right quantity to be able to achieve the ultimate goal of 100% TIP—and it can be done. Overproduction is *bad*. As we run to time, outputs will go up. But what also happens is that rates become less variable. The hour-by-hour output rates stabilize, albeit at a higher level than they were previously. Look back at the chart in the section on rock-busting results at KC. People sometimes miss the significance of the comment: "All machines except B14 demonstrated lower standard deviation for rate of output." As the hourly output rates stabilize, the changeover times improve, and the material supplies move to kanban systems, the difference between time and quantity becomes less. This is shown dramatically in Figure 6.10.

Prior to Lean/R/S, outputs on the line were highly variable, with an average of 2,000 units/hour. After starting Lean/R/S, outputs continued to be highly variable, but the peaks and troughs were getting smaller. The variability in outputs began to fall, and as they did so, output per hour increased. This process continued, resulting in very little variability in outputs, giving stable, consistent, and predictable production at a rate of 2,650 units/hour. This gave them a 33% improvement.

At this point, they could move to a second, more advanced stage of fixed cycles, where the order quantity was fixed. The advantage? Being able to achieve JIT with low stocks of raw materials, packaging supplies, and finished goods.

When output rates are stable and the stage is reached where the switch back to quantity is achieved, people tend to understand why overproduction is bad. So, stopping the lines when the planned quantity is reached seems more sensible than it had been previously. That is how a TIP result of 100% can be achieved. It's a level of performance many would consider impossible, yet we have witnessed a number of companies achieve exactly that after implementing Lean/RfS.

Niggles

Niggle is an English expression that is in popular use in Britain. It does not have a direct translation in many languages and doesn't feature in usage in most English-speaking countries, for example in the United States. So, it's a word a lot of people will *not* be familiar with. A niggle is a small irritation or annoyance. People in different countries have their own word for niggle. At Kimberly-Clark, the machines and equipment are referred to as "assets," so one person with a certain sense of humor at KC renamed niggles as "a pain in the asset," and that's the term KC uses. Others have likened them to small rocks—as in rockbusting—and call them pebbles. In Poland, the name they came up with was "peas." It came from the fairy story of the girl who claimed to be a princess. To test she was telling the truth, she was put in a bed with many mattresses. A pea was placed underneath the bottom mattress. In the morning she complained about the small irritating lump in the bed—the pea. This proved she was a princess, as only a princess would be delicate enough to feel a pea under so many mattresses.

Whatever one calls them doesn't matter. Identifying them and fixing them matters. Fixing niggles makes people less stressed, frustrated, and tired out. It makes them happier, and happier operators, generally speaking, are better operators. One cannot always demonstrate that there will be a direct link between fixing niggles and increased efficiency, but that should not stop any efforts made at trying to eliminate these irritations.

Implementation of the green-stream cycle is usually carried out using a 5-day rapid implementation workshop approach. Niggles is normally one of the subteams comprising around five or six people made up from operators in the area, a mechanic, a manager, and someone not familiar with the area in question. This is the person who can ask "the stupid question," i.e., why things are done certain ways. The first task is to identify what niggles there are. This is done using a check sheet—a simple diagram of the area in question—with the instruction for people to put a cross wherever and whenever a niggle occurs (see Figure 6.11). These check sheets are hung in the area being investigated and left for a few hours, or overnight if there is a night shift. The example shown

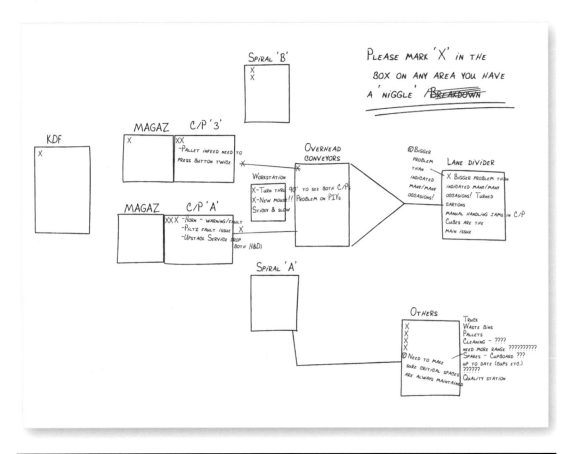

Figure 6.11   Example of a niggles check sheet.

here is of a production line; however it can be any area, including offices or warehouses. It has even been carried out very successfully in the wards and operating rooms of hospitals. The check sheet is an easy way to record niggles; in addition, one can see where they are occurring most often. Note that, in this example, there are boxes for other areas as well as the production line.

Even though people have been asked to just put a cross and not any written explanation, as this would be asking them to do extra work, they often add comments anyway. After leaving the sheets up for a while—certainly less than a day—the subteam returns to ask people to explain why they have placed the crosses where they did. It is not uncommon to find anything from fifty to over one hundred niggles on this kind of sheet. The target for this team is to fix at least 50% by Friday of that week during the rapid implementation workshop—a target that the niggles team usually finds impossible at first. How they do it is up to them. They must also communicate all progress made in the area concerned. One team came up with a clever way to categorize niggles and communicating progress, shown in Figure 6.12.

Each niggle was written on a Post-It® note. The matrix on one axis was difficult/easy to do and the other axis was low/big impact. Color coding was then added to emphasize each category, with green being easy to do with big impact. Red was the opposite, that is, difficult to do with little impact. This display board

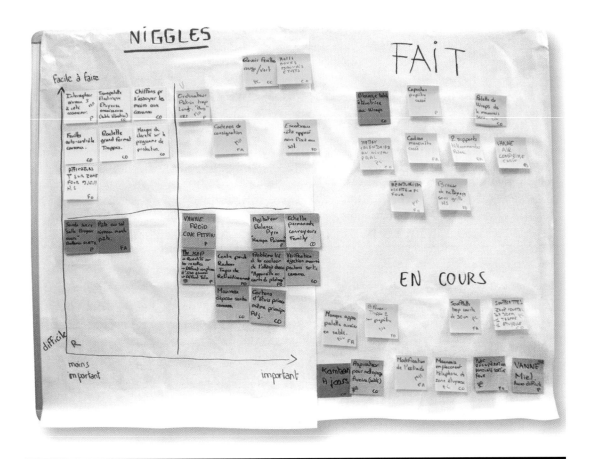

Figure 6.12 Example of a niggles categorization matrix in progress and completed.

was put up in the area where niggles were being tackled ("en cours" = in progress, and "fait" = completed).

In a rapid implementation workshop, the niggles team always has a great week due to the positive feedback they get from people because they have fixed their niggles. They also finish the week exhausted, as they frequently work harder than ever before as they are motivated to fix as many niggles as possible—and they *always* exceed the target of at least 50% that week.

Our experience shows that the concept of *niggles* is easily understood. People *know* about the little irritating annoyances that they endure day after day. Here are a few examples of niggles:

- Nearest phone does not work properly
- Missing basic tools needed to do the job
- Lightbulbs not working
- Leaking valves
- Clutter
- Hard-to-operate taps/valves, etc.
- Inappropriately sized boxes or bins for holding materials
- Points where product jams frequently

■ Points where product can fall off the line
■ Having to walk some distance repeatedly to get materials

The list of niggles goes on and on.

Niggles and the Consultant Surgeon

We ran a 5-day Lean/RƒS rapid implementation workshop at a surgical hospital in the U.K. There were a number of consultant surgeons on the workshop. When the team members were selected for the Monday-afternoon analysis teams, it was decided to put a surgeon in each team. There was, as always, a niggles team to go out and find how many niggles there were.

The surgeon selected to be part of this team was well dressed in an expensive suit and had a rather aloof, even arrogant, attitude. His response to being put in the niggles analysis team was along the lines of "you are not seriously suggesting that I, a consultant surgeon, spend my valuable time finding what niggles exist in this hospital." He was told, "That is exactly what is expected." On Monday afternoon, the niggles team designed their check sheets and then toured the hospital, explaining what they were and what people needed to do. The team decided to leave the check sheets up overnight and come in Tuesday morning to see what crosses had been added. They would then ask the staff about their niggles before reporting back to the rest of the workshop participants on their findings.

To everyone's surprise, it was the consultant surgeon who gave the report back. He plainly spoke from his heart when he said to the assembled,

I had no idea at all of the sort of things people in this hospital are having to put up with. It is outrageous we allow these things to continue. No wonder motivation is so low. I thought it was due to government interference and target setting. It's not—though that might play a part. It is down to us, the senior people in the hospital, not recognizing what nurses, porters, and ancillary staff are dealing with every day. I personally intend to fix as many niggles as I can this week.

He then left the hospital but soon returned dressed not in an expensive suit but in trainers, jeans, and a T-shirt carrying a toolbox. He was good to his word—and the rest of the niggles team did their part too. They worked long hours that week, cajoling and pushing people into helping them get what they needed to fix the niggles. By the end of the week, they had fixed 123 out of a list of 168 niggles—73%. There was a noticeable difference in the surgeon's approach and the way people were interacting with him compared to the Monday morning.

Sometime after the workshop, I received a call from the head of nursing at the hospital. She said the atmosphere at the hospital had completely changed. People were being nicer to each other, helping each other out more, and smiling more. She said the week had changed the culture, and fixing the niggles was what everyone remembered about the week. And that consultant surgeon was no longer seen as aloof or arrogant.

Summary of Chapter 6

There are many Lean tools and techniques to help improve performance. The specific topics covered in more detail in this section are the ones that help to support and sustain the implementation of Lean/RfS and flow logic. However, they are more than that. They are also aimed at making things easier—increasing motivation and reducing the amount of hard work for people. People like to do a good job. They like to set and beat targets. They like to be the best. Fixing the things that cause people stress and extra work will result in improved performance. The real question that needs to be answered is how the company then uses that increased performance.

The next chapter brings all the pieces together. It is about the application of Lean/RfS across all business functions

- To create learning curves, routines, stability, and standards for the most value-adding activities in every business process
- To increase capability and performance in order to outperform key competitors on price, quality, market share, and margins
- To evolve from just Lean/RfS in manufacturing to the point where there is a real Lean advantage in the marketplace
- To beat the competition

Chapter 7

Putting the Pieces Together

If you ask Ian what should come first when "doing Lean," he will always say Lean/R∱S. First, one has to stop the firefighting to create stability as a platform for sustainable continuous improvement. Ask Pascal Dennis, and he would say getting leadership buy-in, understanding, and clear direction. That is strategy deployment and Lean leadership. At Kimberly-Clark, we found, in the light of experience, they were both right, insomuch as you need both. We actually followed Ian's advice, at first, and had some great results. One of the reasons we did this was because, at the time, we really did not understand or appreciate the significance and impact strategy deployment could and would make. We were too focused on wanting to improve manufacturing efficiency. As we rolled out Lean/R∱S across our sites, top management, and other functions, all loved the results. But they really did not understand Lean—they didn't "get it." They saw it as just a set of improvement tools that worked in manufacturing. This increasingly became a problem, as they did not understand what they needed to do to:

1. Help nurture, support, and encourage Lean/R∱S so the improvements continued to grow.
2. Apply the same principles to business processes in other functions.
3. Apply it to the way they personally acted and did their jobs.

Two years after implementing Lean/R∱S, we held the first workshop with Pascal Dennis. In hindsight, KC should have started strategy deployment and Lean leadership shortly after the application of Lean/R∱S in manufacturing, but not before. Why not before? You do need to stop the firefighting that batch logic creates first. Leveled production *is* the foundation of the Toyota production system. It creates the platform that enables sustainable continuous improvement. However, to be able to then make a Lean transformation requires, fairly soon after stability has been achieved, a change in the way the company is led. Leaders in the business need to change the way they manage the business—the way they do their jobs.

Timeline for the Stages in a Lean Transformation

We witness many companies in which their Lean program is focused solely on basic Lean tools and techniques such as value-stream mapping, 5S, TPM, and SMED. They are good improvement tools and techniques. However, without the foundation of Lean/RƒS to break the vicious circle caused by batch logic (see Figure 7.1), they won't deliver the outcomes desired in a Lean transformation. Sustaining improvements is *very* difficult in a firefighting environment.

Lean/RƒS creates a *virtuous* circle that provides stability and economies of repetition where sustainable continuous improvement can flourish. This can be applied to any process, and the improvements can then be invested in faster cycles. These in turn create even more economies-of-repetition benefits (see Figure 7.2).

However, once we start to see these benefits, it is crucial to implement strategy deployment in order to use the results to best advantage against the competition and, perhaps even more importantly, to get senior management to become Lean leaders. Therefore, we now believe the timeline for the stages of a Lean transformation is as illustrated in Figure 7.3.

The length of time in Stage 1 depends on the number of manufacturing locations and product lines in your company. This Stage 1 time could take from a few months to three years or more.

Early aspects of Stage 2 should begin shortly—within weeks—after the first Lean/RƒS workshop in manufacturing. If you want to make a Lean transformation, you must begin the process of implementing strategy deployment and teaching people how to be Lean leaders. If you only want to improve manufacturing efficiencies, then Lean/RƒS will do that for you.

Within Stage 2, the specific activity and timing will depend on many points. The key is to have someone or some part of your organization looking at this

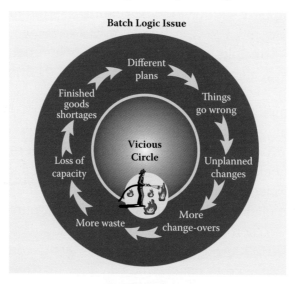

Figure 7.1   Batch logic vicious circle.

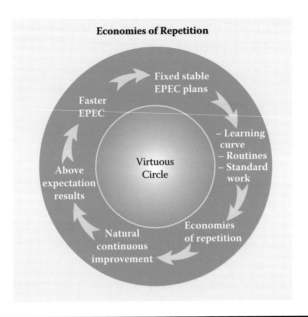

Figure 7.2 Economies of repetition virtuous circle.

roadmap and introducing the right capabilities at the right time. It is our belief that having someone who has already been through Stage 2 to help in this task is essential, that is, someone who has implemented policy deployment and Lean leadership. Otherwise, it's a case of working in the dark. They should also be pushing the leaders of the organization to implement faster than they probably feel comfortable with. For this reason, this person should be an outsider—a Lean sensei.

We have discussed how hard it is to successfully achieve and sustain a Lean transformation and get to Stage 3. Stage 3 is the level where Lean is embedded across the organization and is truly proven as a competitive advantage. Leadership is a key part of this equation. Examples from Jim Womack and Dan Jones's book *Lean Thinking*, including Pat Lancaster from Lantech and John Neill from Unipart, demonstrate that when the leader of the organization understands Lean deeply and then personally applies it to his or her role, this becomes a big differentiator

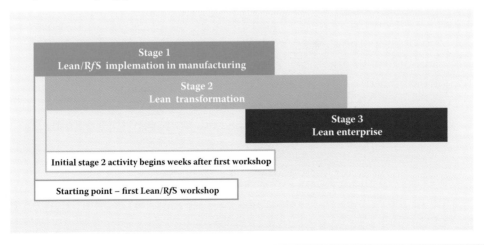

Figure 7.3 Timeline for Lean transformation stages.

in helping become a truly Lean enterprise. We have yet to see an example where a Lean transformation can be pushed to its full potential without it.

Opportunities for Kimberly-Clark

When I first met Ian and saw his "Breaking Through to Flow" presentation several years ago, I made an immediate connection to the flawed logic of batch thinking and knew Lean/R*f*S was a better way. Since then, it has been an incredible journey. We have done projects in one week during Rapid Improvement Workshops that we could not even imagine before Lean/R*f*S. We have applied the principles to improve flow through the supply chain. I have personally learned a great deal about Lean leadership and how to build a problem-solving culture. We have also made mistakes and suffered some setbacks, but continue to go forward on our Lean journey.

Why share all this experience and learning in a book? Surely it should be kept confidential, as it is part of KC's competitive advantage.

The reasons for putting examples and lessons learned from KC into this book are the same reasons why bosses from companies like Lantech and Unipart, mentioned previously, are happy for their experiences to be published. Pat and John see it partly as recognition for their employees. It makes them proud and want to push to do even more. It acts as a motivator and catalyst for stimulating even greater achievements. They also know how difficult the journey was and how long it took to get where they are now. They are further forward in their Lean journey than their competitors. Lastly, they know that to continue on the journey requires cooperation and coordinated effort across all the companies in their supply chains. Suppliers and customers need to act and think in the same way to make the greatest gains possible. These are the reasons for including KC in this book and illustrating where future opportunities lie.

Kimberly-Clark has established Lean/Continuous Improvement as one of our top three enterprise-wide focus areas. This is supported by an organizational commitment with one of KC's key business leaders appointed in the new role of senior vice-president with responsibility for Lean/CI across the company. Tom Falk, our chairman and CEO, announced this role and organization at a forum of several hundred KC leaders. He said:

> My goal is to embed Lean so deep into our company that nothing, no one, could ever pull it out.

That quote is one of the reasons I'm writing this. I intend to help Kimberly-Clark become one of the few global companies to truly succeed with a Lean transformation. At KC, I believe we have that opportunity.

The second reason is what I have seen in my current role as vice-president— customer supply chain. I'm responsible for the flow of product from the point of manufacture to the customer shelf. Prior to this, I had been focusing internally, working with manufacturing and the internal movement of materials. This new role has given me the opportunity to spend time with our customers to work on ways we can improve service to the shelf, while reducing cost and inventory. It has also given me the opportunity to "go see" what really happens to get product into the hands of the consumer.

There are some excellent improvement efforts underway in retail. However, from what I have seen, the opportunity to significantly improve flow and reduce waste in the total value chain is astounding. We have some customers carrying more than fifty days' supply of our products in their stores and distribution centers (DC). Even where the flow and efficiency is better than this, the opportunity to flow without stopping is still big. We have been working on supply chain improvements with one of our key customers for years. We felt the flow and efficiency between KC and retailers was good. Recently, we decided to "go see" with some of our customers. We visited a store, the customer DC, our DC, and a manufacturing plant. It was amazing to see the number of times product was handled, then sat, and then moved again—and again—and yet again before it got to the consumers.

Imagine a green-stream drumbeat rhythm coming from the shelf back to distribution, manufacturing, and raw materials—all working in synchronization to this drumbeat. Think about designing the value stream that ties green-stream flows together while applying the Four Rules of Lean. I believe the benefits would be staggering.

Today, much of the transportation flow is planned based on specific shipments. Usually, each day is a new plan. Yes, there is an understanding of flows at a high level, but these are not set up on a routine repetitive flow that is followed over and over. If we were able to link up customer shipments and supplier shipments from a stable plan of flow in manufacturing, the benefits would be huge, as currently all parts of the supply chain carry additional capacity and inventory to deal with a different plan each time—and one that often changes. We need to bring visibility to the opportunity across the total supply chain if we are to find other companies in KC's supply chain network that will work with us to pursue a Lean transformation on a broader scale.

I want to connect with our supplier and retail partners to tackle this challenge. It is the next big Lean/R*f*S opportunity.

The 5-Day Rapid Implementation Approach

Various sections of this book have described the elements of Lean/R*f*S—but where do we start?

Moving from batch to flow logic is a paradigm shift. At first, it seems illogical, and flies in the face of conventional efficiency wisdom. Trying to get everyone to agree to make the change is difficult, which is why we recommend the 5-Day Rapid

Implementation Approach. This means simply getting the right people together for five days to agree upon what needs to change and to do whatever is required to make those changes. The objective is clear—the implementation of Lean/R*f*S at the site. However, what this looks like at a specific site is up to the attendees to decide. It takes five days to get through all the concerns and to reach agreement between the different functions. Each day, people's understanding of Lean/R*f*S will increase. Each day what seemed ridiculous and impossible before begins to appear more sensible and achievable. This is not something that can be done in a series of meetings spread over weeks or months. There are no project plans or decided actions before the week begins. The participants decide themselves.

Here are the guidelines for typical participants at a 5-day R*f*S workshop. Full week numbers ideally are around twenty-four.

- Five or six team leaders and operators from the area (*must* include actual operators)
- Four or five engineers/technical support
- One or two from commercial area, i.e., sales and marketing
- Three or four managers not necessarily from the area
- Two or three planners (if planning is split between central and local, both areas *must* be represented.)
- No more than eight "outsiders" (e.g., material or equipment supplier if topic relates to what they have or supply, people from sister sites/divisions, or people from other companies, especially if the company is part of a consortium. These "outsiders" can be very helpful during the workshop, as they can take a more objective view.)

These are intended only as guidelines, as each workshop will require its own mix of delegates according to the topic area.

The agenda for the week is not fixed, and what the team objectives will be is determined by the data analysis the participants produce themselves on Monday afternoon and Tuesday morning. Going into a workshop where the tasks to be done during the week are not known beforehand, except for the overall objective that Lean/R*f*S will be implemented, can be a little difficult for some managers to accept.

Here is an overview of the week. The comments on how people typically react each day are in italics.

OVERVIEW OF 5-DAY "BREAKING THROUGH TO FLOW" WORKSHOP

Day 1
- Presentation of Lean/R*f*S concepts
- "Glenday Sieve" followed by VSM of "green" SKUs of current situation

Outcome of Day 1 = understand and appreciate paradigm logic difference of flow, but think it cannot possibly be applied in their business even though "Glenday Sieve" and current-state map shows results exactly in line with what was predicted.

Day 2

- Identification of barriers to changing to Lean/RfS
- Set targets to be achieved by Day 5
- Start rapid implementation aspect of workshop

People see lots of reasons why they cannot do flow—physical, process, and cultural—then identify biggest reasons, as if they can overcome these, and then they can overcome other issues. Targets are what have to be achieved to implement flow—these are always way beyond what they think they can achieve at all, let alone in five days; therefore they still see achieving the paradigm shift as ridiculous.

Day 3

- Continue rapid implementation in teams

Day 3 usually involves some team issues to resolve as well as overcoming some "sacred cows" in the organization—asking "why" five times! Team ends the day on a high as great progress is made.

Day 4

- Team drop into a black hole
- Team reapply PDCA and analysis tools
- Develop and implement breakthrough solutions to the barriers

Teams often hit a "brick wall" on this day as they realize their work is not going to be enough to achieve the targets set; however, they were warned on Day 2 that this would happen—this warning is key to ensuring they come out of the "black hole," as they were also told on day two how to overcome this by going back round PDCA and using brainstorming. This results in them developing breakthrough solutions that achieve above expected results to achieve their targets. Later on in day four, delegates start to see how Lean/RfS can actually be achieved in their business and it suddenly does not look so ridiculous to them. They start to be convinced of the alternative logic of Lean flow and how EPEC (every product every cycle) applies to them.

Day 5

– Finish off rapid implementation, including:

Physical changes

Write up new procedures
Communication and training program for people not involved during the week
Monitoring of ongoing situation
Presentation of results and conclusions of the week

Delegates are totally convinced and converted; however, they also recognize that the rest of the organization now sees them as "mad," so they plan how to convince and convert others through the results they expect to see from implementing Lean/RfS as a practical reality. Attendees fully recognize the extent of the "journey" they have made in five days on changing their paradigm thinking on Lean flow as well as just how much can be achieved in rapid implementation, way beyond what they think is possible.

The focus is on:

■ Using continuous improvement tools and techniques
■ Using data to make decisions on what's required
■ Achieving the target in the time of the workshop
■ *Not* making recommendations
■ Ensuring actions in place to sustain what is achieved (training, monitoring, etc.)

There are usually four or five teams working on different aspects of implementing Lean/RfS. After the first morning, people in each of the teams decide their own start/finish times. This may include early or late finishes (teams have been known to work late into the night!).

Topic areas do not have to be restricted to manufacturing. Rapid Implementation Workshops have been successfully run in warehousing, sales, marketing, product development, and accounts as well as in nonmanufacturing organizations such as hospitals, insurance, and other service companies.

The 5-day rapid implementation approach is typically the method used to get Lean/RfS started at any new site or function. It creates the excitement and energy needed to make that paradigm shift.

During the week, the green team will agree on how Lean/RfS will be integrated into existing planning and management processes. They'll agree on what the Lean/RfS scorecard and KPIs will be. They'll produce communication materials—presentations, handouts, and posters, together with a communication plan to inform everyone impacted by Lean/RƒS. This will state what it is, what is expected of them, and what the benefits will be. They will also agree on the start date—normally one week away to give some time to prepare the communication—and then it goes live. They'll develop the fixed green-stream schedule, buffer tanks, and rules. The driving force throughout is that Lean/RƒS becomes part of how the business is run—not a project, pilot, or yet another new initiative. The implementation of rockbusting, tackling blues and reds, and creating fixed material inbound and outbound flows are all topic areas explored and implemented by the teams during the week.

The rapid implementation week puts all the key Lean/RƒS pieces into place. It is up to the management of the business to ensure that these continue and start to deliver the benefits of economies of repetition and the reduction of firefighting.

Stretching Lean/RƒS

Following a 5-day rapid implementation workshop, economies of repetition usually happen quite quickly, often in a matter of weeks. There is also less firefighting. As a result, people are less stressed and have more time. This gave rise to a problem that was never anticipated. Spotting what opportunities existed to make even more improvements, once flow and stability had been established, had never been a problem for us. However, we found it did not come so easily to others. This was typified by a phone call Ian received from the planning and production managers—Alan and Jackie, respectively—at one company two months after they had implemented their green stream. They called on Monday morning. The conversation went like this:

> "Hi Ian. Alan and Jackie here. Remember how we told you the Monday planning meeting would often last half a day as we looked at last week's plan, what had been made, and what needed to change as a result?"
> "Yes?"
> "Well, today, I asked Jackie what she made last week. And she said exactly the plan. Then she asked me what was the plan for this week, and I said exactly the same as last week. That was it—meeting over in less than a minute."
> "So what's the problem?"
> "What do we do now?"

This came as a complete shock. These were two very capable, intelligent people, yet they just did not know what to do with the time they now had. The suggestion was that they take a walk together from the point materials arrived on site to the point next to the production line where they were ready to be used—a "go see" walk. Then call back. A couple of hours later, the phone rang again.

> "Hi Ian, Alan and Jackie here."
> "How long is the list?"
> "What list?"
> "The list of ideas the two of you have seen for further improvement for material flow."
> "How did you know we had a list?"
> "I know you two and that together you would see opportunities."
> "We do have a list."
> "Well, that's what you now work on."

It had not occurred to us that people would not be able to see for themselves what to work on now that they had more time available. What was needed was some sort of *aide memoire* to help them identify what to work on next. Figure 7.4 shows the model that was developed. We called it "Stretching R/S," subtitled "What do I do now?" It is intended to be used by Lean leaders in discussion with their team to agree on what they should focus on next. It is not in any way a ranking or sequence of activities; rather, it is a menu to choose from as you think appropriate. The key is to always be working on some aspect of improvement when you have achieved stability and repetition from flow logic.

Forces against Flow

We started our first implementation of Lean/R/S at Kimberly-Clark in 2006 in a manufacturing location. We followed the 5-day Rapid Improvement Workshop approach that Ian recommended. It was a great workshop. Teams addressed "barriers to flow" that we identified in the first day. The green-stream team developed a repetitive cycle that produced the green SKUs much more frequently than before. I was on the "blue team" example that was described in Chapter 5. Ian introduced "Niggles" to us in this first workshop, to help find and fix small obstacles that get in the way of the people doing the work from doing it well. KC had many of these. It is a brilliant way to build enthusiasm and excitement on the shop floor. The first workshop was very successful. We came out with a one-week cycle for green-stream production. Years later, the plant remains a great story for improved operating results and working relationships across functions. Since that time, we have run many workshops in plants across the globe. We have extended our Lean thinking and approach in other ways and have learned many things along the way. However, sustaining and building upon this success is wrought with hazards or, as we call them at KC, "forces against flow."

One thing I have learned over the past few years is that the forces against flow are *very powerful*. This section is my warning to you not to underestimate

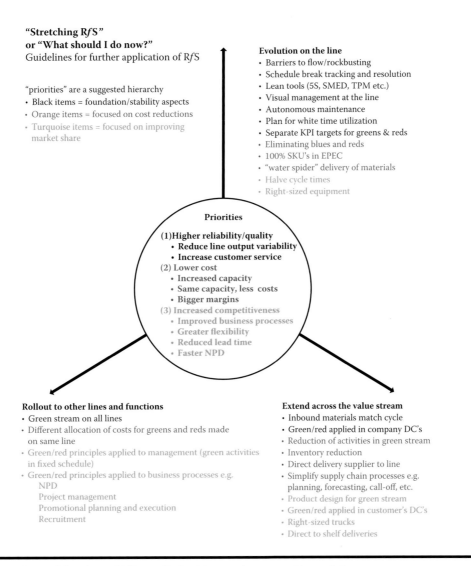

**"Stretching RfS"
or "What should I do now?"**
Guidelines for further application of RfS

"priorities" are a suggested hierarchy
• Black items = foundation/stability aspects
• Orange items = focused on cost reductions
• Turquoise items = focused on improving
  market share

Evolution on the line
• Barriers to flow/rockbusting
• Schedule break tracking and resolution
• Lean tools (5S, SMED, TPM etc.)
• Visual management at the line
• Autonomous maintenance
• Plan for white time utilization
• Separate KPI targets for greens & reds
• Eliminating blues and reds
• 100% SKU's in EPEC
• "water spider" delivery of materials
• Halve cycle times
• Right-sized equipment

Priorities

(1)Higher reliability/quality
  • **Reduce line output variability**
  • **Increase customer service**
(2) Lower cost
  • **Increased capacity**
  • **Same capacity, less costs**
  • **Bigger margins**
(3) Increased competitiveness
  • **Improved business processes**
  • **Greater flexibility**
  • **Reduced lead time**
  • **Faster NPD**

Rollout to other lines and functions
• Green stream on all lines
• Different allocation of costs for greens and reds made
  on same line
• Green/red principles applied to management (green activities
  in fixed schedule)
• Green/red principles applied to business processes e.g.
  NPD
  Project management
  Promotional planning and execution
  Recruitment

Extend across the value stream
• Inbound materials match cycle
• Green/red applied in company DC's
• Reduction of activities in green stream
• Inventory reduction
• Direct delivery supplier to line
• Simplify supply chain processes e.g.
  planning, forecasting, call-off, etc.
• Product design for green stream
• Green/red applied in customer's DC's
• Right-sized trucks
• Direct to shelf deliveries

Figure 7.4   Stretching Lean/RfS—a "what to work on next" model.

these forces. In fact, the more you can understand them and develop a plan to counteract them, the better off you will be. We continue to fight these forces today, and I expect we will for a long time to come.

For some planners, the transition to flow logic can be very difficult. Single-point inventory targets, MRP/ERP, and batch logic are the way they have managed stocks and schedules for years. The vast majority of enterprise and planning software packages are designed on economic order quantity (EOQ) logic. We have had to develop many work-arounds to our systems in order to be able to work with buffer tanks and not replan every week or even every day. As computing power and software capability have increased, planning programs have been created to "optimize the plan" more frequently. In our experience, rather than creating any optimization, these faster programs just create more chaos that we have to fight. I highly recommend getting your IT organization involved early on. Help them to learn and understand flow thinking. Another recommendation—one that many in the IT world will find potentially a little controversial—is to get or retain talent with exceptional skills with Microsoft Excel. They can build spreadsheets capable of running flow logic relatively easily when the improvements focus on the process first and the IT support system second. Have your process fully understood and mapped in detail before trying to develop specific software solutions to support it. *Remember*: Process first; IT systems second.

Get your finance and accounting people involved right up front. We did, but it is still taking time to break through accounting barriers that work against flow. Fixed-asset utilization is a good example of this. For many parts of our business, we operate in a twenty-four-hour, seven-day weekly schedule. KC has benchmarked with many other companies. We have found many industries where the cost of assets is relatively small, and therefore the need for operating at high utilization levels is not too critical. In KC, some of our assets cost more than $10s of millions for a single machine. In this environment, getting a payback on this large capital investment creates the belief that "running full" is a good thing. With Lean/R*f*S, we want to stop the line when the plan is made. This was a *huge* paradigm shift in KC.

The way fixed assets and working capital are accounted for on the P&L and balance sheet work against flow. Key metrics like production rate (units produced per hour) need to be replaced with production conformance (producing exactly to plan). Balancing the cost of inventory versus "operational efficiency" needs to be considered differently. Then there is product costing to consider—separating the actual efficiencies and changeover costs of producing green SKUs from red ones—and then reflecting these different figures in the product costing.

Accounting people were against this at first, mainly because they did not believe manufacturing could consistently keep to the same fixed plan each week. They were concerned that they would not be able to monitor the green and red SKUs separately. Once we were able to show that it could be done, they became much keener on the idea of costing greens and reds separately. They, like everyone else, always knew that the reds cost more to produce than traditional product costing implied, but they were not able to prove it when the plan was different every time. With Lean/R*f*S, they could prove it. It enabled them to help marketing and sales make more informed decisions on how to treat the reds, based on a more accurately calculated profit margin. Or, as happened with many reds, what was now a *negative* margin! Of course, the margins on the greens looked a lot better. This also led to different decisions being taken on how these products were marketed and what terms to give retailers, with the objective of increasing market share at the same time as overall improved profit margins.

Lean/R*f*S Fundamental Beliefs

After many years of applying Lean/R*f*S, we have come to some fundamental beliefs about Lean/R*f*S that are "required" to achieve a true Lean transformation with sustainable processes and breakthrough results.

Lean/RfS is a journey. Although immediate results from Rapid Improvement Workshops can and will definitely be demonstrated, results that are truly breakthrough will take years. There are definitely good demonstrated improvements in the early years of Lean/RfS. However, what is noticeable is that at about Year 5 there is a big inflection point where results can ramp up significantly. At KC, we believe it is the combination of strategy deployment, Lean leadership, and stability in the supply chain all coming to a point of maturity that creates a tipping point in performance.

Lead time needs to become a critical measure. The pursuit of lower lead time should be a never-ending goal. Time is a resource that cannot be regenerated. Once it is past, it is gone. This is also a true measure of how Lean, waste-free, and flowing processes are performing. Lead time is a measure that is not sufficiently understood or valued. In the companies we have seen that are most advanced in Lean, reducing lead time is always a key priority. Taking time out of the process directly correlates to how well your process is flowing. There is a story told about Taiichi Ohno. Whenever anyone came to him with an idea for an improvement, his response was always the same: "Will it reduce lead time?" If the answer was no, then his response was, "I'm not interested." To him, lead-time reduction was king, as he believed that the company that converted orders to cash the fastest would be the best. The products we make are just the physical objects used to produce the real product of a manufacturing company—*money*. So the quicker orders are converted into cash, the better the company is.

Pursue small quantities, more frequently, for virtually everything. It is easy to find examples and stories where we have applied this principle in a production environment. It is often hard for most people to connect this to other activities. The belief that production of small quantities, produced more frequently, is the right thing to do is counterintuitive. Only repeated experiences that it does work helps build understanding and capability to a level where team members start to see and do this automatically. At KC, we had a recent example where we wanted to address a gap in our DC inventory accuracy. Our standard is very high, 99.5%, but with the volume of shipments in our process, even a very small error can cause a negative impact for our customers. This measure was traditionally reviewed on a quarterly basis. Imagine how hard it was to understand what was causing inventory inaccuracy over this period of time. The information was in a big "batch" of three months of data. The team moved the measure to monthly. This helped, but it still did not provide sufficient information to understand the root causes of inventory inaccuracy. Then the team developed an approach where any DC that was below the monthly standard would be checked and reviewed daily for any inventory discrepancies. The accountability of being checked every day and looking at problems daily had an immediate impact. The pace of improvement from there was astounding.

Stop overproduction. It is the worst form of waste. It causes all sorts of other waste. For KC, with high-cost assets, high asset utilization was seen as good. The forces that encouraged overproduction to happen were very powerful. The old metrics and cultural norms made it no big deal to overproduce. They even rewarded this behavior. When the plan is made, stop producing. This is a simple message, but one that people find very hard to accept.

Despise inventory. We are not saying eliminating all inventories should be the goal. It is a necessary component to buffer against variability. However, getting to a point where you are always asking, *"Why* do we have this inventory?" should be the norm. A strong ongoing focus on reducing inventories is essential. It demonstrates greater capability, reduces numerous wastes, releases cash, and lowers lead times.

Summary of Chapter 7

Anyone who likes to do jigsaw puzzles knows you put the corner pieces in first and then add the straight edges before filling in the center pieces. We have shown you what we believe each of these Lean/R∕S pieces is and, therefore, in what sequence they should be put together. A key lesson from KC—and other companies—is that all the edge pieces need to start very soon after implementing the first steps of leveling. This particularly applies to the edge pieces of Lean leadership and policy deployment, that is, if you want to achieve a Lean transformation in your company. If you only want to improve your manufacturing efficiencies, then just Lean/R∕S will do that for you. What we are certain of is that implementing Lean tools in an environment where every plan is different—and is then often changed—is *never* going to create a sustainable continuous improvement culture. This is the problem created by batch logic—the logic used in most ERP/MRP planning systems.

We hope you have found this book helpful and a spur to drive forward a Lean transformation with increased momentum and understanding in your company. Good luck.

Ian Glenday and Rick Sather

Glossary

5S: A Lean improvement technique of activities aimed at promoting organization and efficiency in the workplace

Annualized hours: A payment system where total required hours in the year is calculated and paid in equal weekly amounts even though actual working hours per week may vary

Breakthrough: An improvement or change to the way things are done that takes performance to a level previously not thought possible

Center-lining: Creating a standard datum point in a machine from which standard settings can be established

EPEC: Every Product Every Cycle

FMCG (Fast-Moving Consumer Goods): Products people buy frequently, e.g., grocery goods

Gemba: Japanese for shop floor in a factory

Gemba walk: A walk through the factory looking for further opportunities to improve

JIT (just-in-time): Deliveries of materials from suppliers that arrive just before they are required in production

KPI: Key Performance Indicator

Muda: Japanese word for waste

Ossified: Thickening of bone, especially around a joint, that reduces flexibility and mobility

Passive resister: A person who, through his or her actions, shows that he or she is subconsciously against what is being implemented

Re-sieve: Process, usually two weeks before the end of a green stream cycle, when the next green cycle schedule is agreed

Run full: A schedule that loads a machine to run twenty-four hours a day, seven days a week

Sensei: Japanese term for an experienced coach or mentor

Side bar: A Kimberly-Clark term for a quickly arranged meeting with senior site management during a 5-day workshop to decide whether to proceed with a breakthrough opportunity or not

SKU (Stock Keeping Unit): A product or item that is sold to consumers

SMED (Single Minute Exchange Die): A Lean improvement technique aimed at reducing change over times

TPM (Total Preventative Maintenance): A Lean technique aimed at improving efficiency through an effective maintenance program

TPS (Toyota Production System): The way that Toyota operates; the basis of Lean

True north: Expresses a business need that must be achieved to continue to be successful

VSM (Value Stream Mapping): Lean technique to visualize and quantify the supply chain

White space: Measure of improvements achieved through economies of repetition. When running to time, this equates to more output in the same time. When running to quantity, it is time, as production produces the same amount in less time.

Recommended Reading

Dennis, P. (2006). *Getting the right things done*. Cambridge, MA: Lean Enterprise Institute.
Glenday, I. F. (2005). *Breaking through to flow*. Ross-on-Wye, UK: Lean Enterprise Academy.
Womack, J. P., & Jones, D. T. (1996). *Lean thinking*. New York, NY: Simon & Schuster.
Womack, J. P. (2011). *Gemba walks*. Cambridge, MA: Lean Enterprise Institute.

Appendix A

Lean/R *f*S scorecard example

		Score card - RfS Line 8	
		period 1	
	TARGET	W11	W12
Process manufacturing KPIs			
TOP (green from w41)	100%	107.2%	103%
TIP	98%	92.80%	97.00%
Schedule breaks	0	1	0
Green cycle breaking	0	0	0
Timing conformance green (minute)	02.00	2.26	4.13
Operators workshop meetings	1	0	1
Process supply chain KPIs			
Stock outside max. limits (number) - domestic	0	0	0
Stock outside min. limits (number) domestic	0	0	0
Number of greens produced outside green stream	0	3	0
Production of Greens outside green stream in %	0%	8%	0%
Results KPIs			
Operational efficiency	92%	93.8%	95.6%
Production efficiency	95%	98.0%	99.8%
Waste	0.8%	1.1%	0.6%
Material shortages (no.of time)	0	0	0
% of greens in total # of SKU	15%	15%	15%
% of greens in volume	65%	68%	68%

Figure A1.1 Lean/R*f*S scorecard example.

Lean/RfS scorecard definitions

TOP	TOP = (delivered qty - target qty)/target qty
TIP Plant service level (quantitative)	PSL = absolute (delivered qty - target qty)/target qty
Schedule breaks	Number of plan changes (red and green)
Green cycle breaking	Number of plan changes in green stream
Timing conformance	Σ absolute (starting actual time for a batch - planned starting time for a batch)/number of batches
Stock outside limits	How many SKUs have an inventory outside buffer
Operational efficiency	net production time/operational time %
Production efficiency	net production time/production time %
Waste	Waste = time lost due to waste production/production time %
Number of greens outside green stream	Number of "greens" SKU produced outside green stream
Greens outside green stream %	greens produced outside green stream/greens produced in green stream %

Figure A1.2   Scorecard definitions.

Index